Gary Smith has done it again. *Distrust* is a wild ride that derails the Big Data hype train with force, style, and above all sardonic humour. Smith is a master of illustrating by example—examples that are fresh, unexpected, at times shocking, and at times hilarious. Come along on Smith's tour of statistical snake-oil and you'll never look at AI or data science the same way again.

—Carl T. Bergstrom
Professor of Biology, University of Washington
Author of *Calling Bullshit: The Art of Skepticism in a Digital World*

****

Any fan of Carl Sagan's *The Demon Haunted World* will love this book. Like Sagan, Smith discusses the challenges to human progress that result from a lack of critical thinking skills, and he does so with a Sagan-esque keen eye and eloquent voice. Smith also makes clear how the threats to sound judgment and effective decisions are more formidable than those of Sagan's day, as faulty thinking is now aided and abetted by an Internet-fuelled distrust of science, viral misinformation, and venomous conspiracy theories. The wisdom in this book is desperately needed.

—Tom Gilovich
Irene Blecker Rosenfeld Professor of Psychology
Cornell University
Author of *The Wisest One in the Room*

****

It turns out that, unlike the mythical hero, AI has two Achilles' heels. Not only are the technologies not intelligent, more perniciously, neither are too much of the statistics and data use on which AI and big data rely. Gary Smith provides a brilliantly executed counter against pseudo-science and the accumulating garbage we misleadingly call information, including timely and important warnings and ways forward for policy-makers, practitioners, academics, and citizens alike.

—Leslie Willcocks
Professor Emeritus, London School of Economics and Political Science

****

An immensely readable look at why we need science more than ever, but also why and how science needs to clean up its act. Recommended for anyone who occasionally wonders whether that "outspoken" family member on Facebook might just have a point.

—Nick Brown, PhD, scientific integrity researcher

****

Smith marvellously illustrates the evolution of disinformation. He richly demonstrates how blind faith in technology enables more misrepresentations of the truth. *Distrust* articulates a humbling view of how we should think critically about new findings from hyped technology trends.

—Karl Meyer, Managing Director, Digital Alpha Advisors LLC
Former Partner at Kleiner Perkins

****

Distrust

DISTRUST

Big Data, Data Torturing, and the Assault on Science

GARY SMITH

OXFORD
UNIVERSITY PRESS

OXFORD
UNIVERSITY PRESS

Great Clarendon Street, Oxford, OX2 6DP,
United Kingdom

Oxford University Press is a department of the University of Oxford.
It furthers the University's objective of excellence in research, scholarship,
and education by publishing worldwide. Oxford is a registered trade mark of
Oxford University Press in the UK and in certain other countries

Published in the United States of America by Oxford University Press
198 Madison Avenue, New York, NY 10016, United States of America

British Library Cataloguing in Publication Data
Data available

Library of Congress Control Number: 2022940620

ISBN 978–0–19–286845–9

DOI: 10.1093/oso/9780192868459.001.0001

Printed and bound by
Printed by Integrated Books International, United States of America

Links to third party websites are provided by Oxford in good faith and
for information only. Oxford disclaims any responsibility for the materials
contained in any third party website referenced in this work.

CONTENTS

PART V THE CRISIS

INTRODUCTION

Disinformation, Data Torturing, and Data Mining

W hat were the greatest inventions of all time—not counting fire, the wheel, and sliced bread? The printing press has to be near the top of anyone's list because it allowed ideas and inventions to be spread widely, debated, and improved. Electricity, internal combustion engines, telephones, vaccinations, and anesthesias are also worthy candidates. These are just the very, very top of a very, very long list. Look around you, at your computer, eyeglasses, refrigerator, plumbing, and all of the other everyday things we take for granted.

Where would we be without science—living short brutish lives in caves? I exaggerate, but not much, and only to make the point that science and scientists have enriched our lives beyond measure. I include economists, psychologists, and other "softer" scientists in this group. During the Great Depression in the 1930s, governments everywhere had so little understanding of the economy that their policies were utterly counterproductive. When Franklin Roosevelt campaigned for U.S. President in 1932, he promised to reduce government spending by 25 percent. One of the most respected financial leaders, Bernard Baruch, advised Roosevelt to "Cut government spending—cut it as rations are cut in a siege. Tax—tax everybody for everything." The Federal Reserve let the money supply fall by a third and Congress started a trade war that caused exports and imports to fall by more than 50 percent.

Today, thanks to theoretical and empirical economic models, we (well, most of us) know that spending cuts, tax hikes, monetary contractions, and trade wars are exactly the wrong policies for fighting economic recessions. This is why, during the global economic crisis that began in the United States in 2007, governments did not repeat the errors of the 1930s. With the world economy on the brink of a second Great Depression, governments did the right thing by pumping trillions of dollars into their deflating economies. Ditto with the COVID-19 economic collapse in 2020. Plus, scientists developed the vaccines that tamed the COVID pandemic.

Unfortunately, science is currently under attack and scientists are losing credibility. There are three prongs to this assault on science:

Disinformation

Data torturing

Data mining

Ironically, science's hard-won reputation is being undermined by tools invented by scientists. Disinformation is spread by the Internet that scientists created. Data torturing is driven by scientists' insistence on empirical evidence. Data mining is fueled by the big data and powerful computers that scientists created.

In this Introduction, I will use examples related to bitcoin to illustrate this tragedy.

Money

Money is a secret sauce for economic progress because it allows people to specialize in jobs they are good at and use the money they are paid to buy things from people who are good at other jobs. Pretty much anything can serve as money as long as people believe that it can be used to buy things. If that sounds circular, it is.

In practice, over the past 4000 years, the predominant moneys have been precious metals—mostly silver, to a lesser extent gold, and even less frequently copper. However, an enormous variety of other moneys have also been used. Wampum (beads made from shells) was legal tender in several American colonies. Tobacco was used as money in Virginia for nearly 200 years, in Maryland for 150 years, and in neighboring states for shorter periods. In fact, a 1641 law made tobacco legal tender in Virginia and prohibited contracts payable in gold or silver. Rice, cattle, and whiskey have been legal tender and, although they lacked legal sanction, musket balls, hemp, furs, and woodpecker scalps have also been used as money.

The island of Yap in Micronesia is renowned for using stone money for more than 2000 years. As shown in Figure I.1, these *rai stones* are round like a wheel with a hole in the center and often taller than a person. The stones were obtained by canoe expeditions to Palau or, for finer and rarer stones, to Guam some 400 miles away. The seas were often stormy and the destinations hostile. Many canoes never returned.

Figure I.1 A rai stone.

In the 1870s, David Dean O'Keefe, an American with a sturdy boat, brought enormous stones into Yap in exchange for coconuts, fish, and whatever else he wanted. The largest stone, said to be 20 feet tall, is at the bottom of the Yap harbor, where it fell while being unloaded from his schooner onto a raft. Even though the stone disappeared from view, it is still considered to be money by generation after generation of owners.

Today, the islanders use stone beads, sea shells, beer, and U.S. dollars for small transactions. The large stones are used to pay for land, permission to marry, political deals, and substantial debts. When a large stone changes ownership, a tree can be put through its hole so that it can be rolled from its old owner to its new one. Because a broken stone is worthless, stones are often left in one spot, with the ownership common knowledge. The largest stones are so well known that they have names. The people pass on, but the stones remain.

The stones acquired before the arrival of O'Keefe are the most valuable; those brought in by O'Keefe are worth half as much; and more recent stones are essentially worthless. In 1984 *The Wall Street Journal* reported that an islander bought a building lot with a 30-inch stone, explaining that "We don't know the value of the U.S. dollar."

Using stones for currency seems odd, but is it any odder than using paper money? The Yap islanders accept rai stones as payment because they are confident that they can use the stones to buy things they want. Their acceptance as money rests on this confidence, nothing more. The same is true of our money. It is worthless except for the fact that it can be used to buy things.

Money doesn't need to be something we can hold in our hands, and most of us choose to live in an essentially cashless society using electronic records of transactions. Our income can go directly into our bank accounts and our bills can be paid with a check, debit card, credit card, or directly from our bank accounts. Cash is only needed for paying people who won't take checks or credit cards—perhaps for buying food at a local farmer's market, paying our children to mow the lawn, or making purchases we don't want the government to know about.

Even so, the amount of cash in circulation continues to grow. There is now more than $2 trillion in U.S. currency in circulation, including 18 billion $100 bills—an average of 54 Benjamins for every American man, woman, and child. There are more $100 bills than $1 bills.

Where are these 18 billion Benjamins? It has been estimated that more than half of all U.S. cash is held outside the United States. Some is held as reserves by central banks; some is used by citizens and businesses as a second currency or as a hedge against economic instability. Much of the cash, domestically and abroad, is used for illegal or untaxed activities that leave no electronic record.

Which brings us to bitcoin, the original and most well-known cryptocurrency. The bitcoin concept was unveiled in a 2008 paper written by someone using the pseudonym Satoshi Nakamoto and was implemented the following year.

When news stories are written about bitcoin, there is often, as in Figure I.2, an accompanying image of a gold (or sometimes silver) coin with a large B and vertical hashmarks, like a dollar sign.

Figure I.2 A mythical bitcoin.

There is, of course, no bitcoin coin. Bitcoins and other cryptocurrencies are digital records—like bank accounts and credit cards. The difference is that bank account and credit card records are maintained by financial institutions and tied to social security numbers, while cryptocurrency transactions are recorded in a decentralized blockchain that, in theory, preserves the anonymity of users.

Nobody with a credit card or a checking account needs to use $100 bills or bitcoins, but people who want to buy or sell things in secret like them. It should not be surprising that governments have been able to force bitcoin dealers to turn over financial records, leading to arrests and bitcoin confiscations. Unlike bags of $100 bills, blockchain technology can actually create an electronic trail that helps law enforcement officials follow the money and document criminal transactions.

Blockchain transactions are so slow that the value of a bitcoin can change substantially between the beginning and end of a transaction. Transactions are also expensive, and environmentally unfriendly. In 2021, Cambridge University researchers estimated that bitcoin blockchains consume as much electricity as the entire country of Argentina, with only 29 countries using more electricity than bitcoin. Bitcoin is essentially strip-mining the climate.

In June 2022, a group of 1500 prominent computer scientists signed a letter to U.S. Congressional leaders stating that blockchain technology is "poorly suited for just about every purpose currently touted as a present or potential source of public benefit."

The fact that bitcoin has no inherent value makes it a great example of the three themes running though this book: disinformation, data torturing, and data mining.

Disinformation

A large part of the initial allure of bitcoin was that the underlying blockchain technology was not created by the economic and financial elite but by a mysterious outsider whose identity is still unknown. A David stepped up to battle Goliath. Bitcoin is not controlled, regulated, or monitored by governments or the banking system. *The Declaration of Bitcoin's Independence*, endorsed by numerous celebrities, states that

We hold these truths to be self-evident. We have been cyclically betrayed, lied to, stolen from, extorted from, taxed, monopolized, spied on, inspected, assessed, authorized, registered, deceived, and reformed. We have been economically disarmed, disabled, held hostage, impoverished, enervated, exhausted, and enslaved. And then there was bitcoin.

Paul Krugman, an economics Nobel Laureate and *New York Times* columnist, has written that bitcoin is

something of a cult, whose initiates are given to paranoid fantasies about evil governments stealing all their money . . . Journalists who write skeptically about Bitcoin tell me that no other subject generates as much hate mail.

I learned firsthand about the cult's paranoia when I wrote some *MarketWatch* columns about bitcoin. Here are a few responses (with spelling mistakes corrected) from infuriated readers:

Rich people fear Bitcoin as it means their wealth could be shifted over to "peasants" and are scared of becoming irrelevant.

Bitcoin makes cross border payments possible and provides an easy way for people to escape failed government monetary policy.

Nobody believe this orchestrated FUD which is all paid for by American Express and other banking institutions.

FUD is a popular acronym for "fear, uncertainty, and doubt." Another is HODL. An enthusiast once misspelled *HOLD* as *HODL* (there are lots of misspelling on bitcoin forums) and it was misinterpreted as an acronym for "hold on for dear life." Bitcoin fanatics ignore the FUD and just HODL.

The irony here is that scientists created our highly efficient financial system. They also created the Internet that broadcasts the bitcoin gospel that denounces our financial system. Conspiracy theories and paranoia are as old as the human race, but the Internet has made their circulation fast and frenzied.

Tortured Data

Cryptocurrencies are a lousy way to buy and sell things. On the other hand, many gullible speculators think that wildly fluctuating bitcoin prices will make them rich. Starting from a price of $0.0008 in 2009, bitcoin's price topped $67,000 in November 2021. Figure I.3 shows the wild ups and downs in prices since 2014.

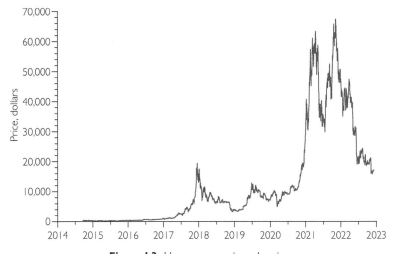

Figure I.3 Here we go again and again.

It is no surprise and that professionals and amateurs have searched for ways to predict which direction bitcoin prices will go next. The unsolvable problem is that there is no rational explanation for movements in bitcoin prices. How can anyone predict the irrational?

The *intrinsic value* of an investment is how much you would be willing to pay to hold it forever. The intrinsic value of a stock is the price you would pay to get the dividends. The intrinsic value of a bond is the price you would pay to get the interest. The intrinsic value of an apartment building is the price you would pay to get the rents. The intrinsic value of bitcoin is zero because you get no cash from holding it.

Investors would buy stocks, bonds, and apartment buildings even if there were a law that prohibited them from ever selling. They would not buy bitcoins under that condition because the only way to make a profit is to sell them to someone else. This is called the Greater Fool Theory—buy at a foolish price, hoping to sell to an even bigger fool.

During the Dutch tulip bulb bubble, bulbs that might have fetched $20 (in today's dollars) in the summer of 1636 were bought by fools in January for $160 and bought by bigger fools a few weeks later for $2000. The prices of exotic bulbs topped $75,000. During the South Sea Bubble in the 1700s, fools bought worthless stock in hopes of selling to bigger fools. One company was formed "for carrying on an undertaking of great advantage, but nobody is to know what it is." The shameless promoter sold all the stock in less than five hours and left England, never to return. Another stock offer was for the "nitvender" or selling of nothing. Yet, nitwits bought nitvenders.

Bitcoins are a modern-day nitvender. Now, as then, prices are driven by fear, greed, and other human emotions—what the great British economist John Maynard Keynes called "animal spirits." People buy bitcoin because they hope that bigger fools will pay higher prices.

In 2017, as the bitcoin bubble picked up speed, the stock price of Long Island Iced Tea Corp. increased by 500 percent after it changed its name to Long Blockchain Corp. At the peak of the bitcoin bubble, a company introduced a cryptocurrency that didn't even pretend to be a viable currency. It was truthfully marketed as a digital token that had "no purpose." Yet, people spent hundreds of millions of dollars buying this nitvender.

The Wild West of Banking, Again

On July 13, 2022, the CoinMarketCap website reported the prices of 9909 cryptocurrencies with a total market value of $900 billion. Seeing this, I was reminded of the Wild West period of American banking. In the early 1800s, there were thousands of bank-issued currencies. One historian wrote that

corporations and tradesmen issued "currency." Even barbers and bartenders competed with banks in this respect . . . nearly every citizen regarded it as his constitutional right to issue money.

One successful Midwestern banker recalled his start in the business:

Well, I didn't have much to do and so I rented an empty store and painted "bank" on the window. The first day a man came in and deposited $100, and a couple of days later, another man deposited another $250 and so along about the fourth day I got confidence enough in the bank to put in $1.00 myself.

By the time of the Civil War, there were 7000 different bank currencies in circulation, of which 5000 were counterfeit. Banks were state-regulated and regulations regarding precious-metal reserves were lax. A Massachusetts bank that had issued $500,000 in bank notes had $86.48 in reserves. In Michigan, a common collection of reserves (including hidden lead, glass, and ten-penny nails) passed from bank to bank, ahead of the state examiners. Opening a bank and issuing bank notes seemed like an easy way to literally make money. It was an occupation that attracted the most reputable and public-spirited people and the lowest and most down-and-out scoundrels. Many helped the country prosper; others simply redistributed its wealth.

Figure I.4 shows the lavishly embellished $1 bill issued by the Detroit City Bank, which was said to be backed by "Pledged Property" of "Real Estate & Private Holdings." The bank opened in 1837 and closed in 1839.

Figure I.4 A $1 bill issued by the Detroit City Bank.
Credit: Paper Money Guarantee.

There was no need for thousands of bank currencies in the nineteenth century and there is no need for thousands of cryptocurrencies today. Perhaps the most ludicrous (it is admittedly a tough call) is Dogecoin, which was created as a joke by two software engineers. It advertises itself as the "fun and friendly internet currency" and uses the face of a Shiba Inu dog from the "Doge" meme as its logo (Figure I.5). In the first 30 days after its official launch on December 6, 2013, more than a million people visited dogecoin.com. In January 2014, Dogecoin's trading volume was larger than that of all other cryptocurrencies combined. Soon, every Dogecoin tweet from Elon Musk sent shockwaves to Dogecoin prices. On May 7, 2021, Dodgecoin's total market value hit $82 billion, making it the third most valuable cryptocurrency (behind bitcoin and Ethereum). Two months later, its value had plummeted 90 percent and Musk was sued for market manipulation.

Some struggled to make sense of this nonsense. An economist at the University of New South Wales Business School described Dogecoin as "not so much an alternative deflationary numismatic instrument as it is an inflationary leisured exploration of community-building around a cryptoasset." That kind of gobbledegook hardly enhances the reputation of scientists.

Meanwhile, the two Dogecoin co-founders bailed, evidently exasperated and embarrassed that their joke had gotten out of control. One wrote that, "Cryptocurrency is a solution in search of a problem . . . . If anything, it exists as an educational tool. It's a reminder that we can't take this stuff seriously." The other explained his departure: "I was like, 'Okay, this is dumb. I don't want to be the leader of a cult.'"

Figure I.5 A cute puppy coin.

Just When You Thought it Couldn't Get Any Crazier

As if cryptocurrencies weren't sufficiently foolish, we now have CryptoKitties and other crypto-collectibles. Bitcoins are *fungible* in that all bitcoins are identical and interchangeable. A crypto-collectible, in contrast, is a unique, non-fungible digital token (NFT) that cannot be copied, substituted, or subdivided.

Almost all crypto-collectibles are traded and validated on Ethereum's blockchain network. CryptoKitties were unveiled in 2017 and prices soared, with the Dragon cryptoKitty selling for $172,000 in 2018. Cryptopunks were unleashed in 2017, and as of December 9, 2022, the creators of cryptopunks reported that there had been 15,496 sales for a total of $2.47 billion. In June 2021, the Alien Cryptopunk with a Mask shown in Figure I.6 sold for $11.8 million to Shalom Meckenzie, the largest shareholder of DraftKings, a fantasy sports betting organization.

If it seems fitting that a fantasy sports mogul would buy fantasy art, it is even more ironic that in 2018, a group claiming to be bored MIT students created EtherTulips—NFTs that look like tulips. The founders were brutally honest:

[L]ots of people have been calling crypto the biggest bubble since Tulip mania. Well, we're a group of bored college students and decided to push the craze to the next level by putting virtual tulips on the Ethereum blockchain!

The Ethertulips website is equally candid:

Tulip mania was a bubble in the 17th century during which prices for tulips rose to ridiculous levels. One bulb cost more than a house! The tulip market crashed in 1637, so you were born too late to participate—but no worries! We're bringing tulip mania to the 21st century.

I was convinced that these were indeed bored college students when they included the punchline: "What could go wrong?"

Figure I.6 The Alien Cryptopunk with a mask.

Credit: Courtesy of Sotheby's and Larva Labs.

Market Manipulation

There is another important factor driving cryptocurrency prices. In a pump-and-dump scheme, a group of scammers circulate untruthful rumors about an investment, while trading it back and forth among themselves at ever higher prices, luring in the credulous. After prices have been pumped up, the conspirators dump their holdings by selling to the suckers (aka "bag holders"). In 2019, *The New York Times* reported that Donald Trump had been involved in a handful of pump-and-dump schemes in the late 1980s:

As losses from his core enterprises mounted, Mr. Trump took on a new public role, trading on his business-titan brand to present himself as a corporate raider. He would acquire shares in a company with borrowed money, suggest publicly that he was contemplating buying enough to become a majority owner, then quietly sell on the resulting rise in the stock price.

The tactic worked for a brief period—earning Mr. Trump millions of dollars in gains—until investors realized that he would not follow through.

Bitcoin's price surges often seem like pump-and-dump scams. Krugman has written that

Bitcoin's untethered nature also makes it highly susceptible to market manipulation. Back in 2013 fraudulent activities by a single trader appear to have caused a sevenfold increase in Bitcoin's price. Who's driving the price now? Nobody knows. Some observers think North Korea may be involved.

In 2019 the *Wall Street Journal* reported that nearly 95 percent of the reported bitcoin trades are fake trades intended to manipulate prices. A 2020 study published in the *Journal of Finance* concluded that nearly all of the rise in bitcoin prices in 2017 was due to trading by one large, unidentified trader using another digital currency, called Tether, to buy bitcoin.

In a 2021 report, Research Affiliates, a widely respected investment management company, concluded that

perhaps [bitcoin] is just a bubble driven by a frenzy of retail, and some institutional, money eager to get a piece of the action. Alternatively, and far likelier in my opinion, is that this "bubble" is more fraud than frenzy.

On August 3, 2021, the head of the Securities and Exchange Commission (SEC) said that cryptocurrency markets were "rife with fraud, scams and abuse." In June 2022 the U.S. Department of Justice charged six individuals with cryptocurrency fraud. As I write this, a lot of people are waiting for more shoes to drop.

Figure I.3 shows the run-up in bitcoin's price in 2017, culminating in a peak price of $19,497 on December 16, followed by sharp decline. There was no reason for this up and down other than investor speculation and/or market manipulation. It happened again on Monday, April 1, 2019, as the volume of bitcoin trading more than doubled and the price jumped 17 percent. The rally was concentrated in a one-hour interval when the price leapt 21 percent between 5:30 a.m. and 6:22 a.m. London time. Who let the fools out? One plausible explanation was an article written as an April Fool's joke reporting that the SEC had held an emergency meeting over the weekend and voted to approve two bitcoin-based exchange traded funds (ETFs). For true fools, it hardly mattered what sparked the rally. As long as prices are going up, fools will buy in hopes of selling to bigger fools.

After a lull, bitcoin price and volume took off again in late October 2020. When it will end, no one knows. But it will end—badly. At some point, the supply of greater fools will dry up, the manipulators will dump their bitcoin, and the bitcoin bubble will end the way all bubbles end. Today, we laugh at the Dutch who paid the price of a house for a tulip bulb. Future generations will laugh at us for paying the price of a fancy car for literally nothing.

Torturing Data to Predict Bitcoin Prices

The fact that changes in bitcoin prices are driven by fear, greed, and manipulation has not stopped people from trying to crack the secret of bitcoin prices. Empirical models of bitcoin prices are a wonderful example of data torturing because bitcoins have no intrinsic value and, so, cannot be explained credibly by economic data.

Undaunted by this reality, a National Bureau of Economic Research (NBER) paper reported the mind-boggling efforts made by Yale University economics professor Aleh Tsyvinski and a graduate student, Yukun Liu, to find empirical patterns in bitcoin prices.

Tsyvinski currently holds an endowed chair named after Arthur M. Okun, who had been a professor at Yale from 1961 to 1969, though he spent six of those eight years on leave so that he could work in Washington on the Council of Economic Advisors as a staff economist, council member, and then chair, advising presidents John F. Kennedy and Lyndon Johnson on their economic policies. He is most well known for Okun's law, which states that a 1 percentage-point reduction in unemployment will increase U.S. output by roughly 2 percent, an argument that helped persuade President Kennedy that using tax cuts to reduce unemployment from 7 to 4 percent would have an enormous economic payoff.

After Okun's death, an anonymous donor endowed a lecture series at Yale named after Okun, explaining that

Arthur Okun combined his special gifts as an analytical and theoretical economist with his great concern for the well-being of his fellow citizens into a thoughtful, pragmatic, and sustaining contribution to his nation's public policy.

The contrast between Okun's focus on meaningful economic policies and Tsyvinski's far-fetched bitcoin calculations is striking.

Liu and Tsyvinski report correlations between the number of weekly Google searches for the word *bitcoin* (compared to the average over the past four weeks) and the percentage changes in bitcoin prices one to seven weeks later. They also looked at the correlation between the weekly ratio of *bitcoin hack* searches to *bitcoin* searches and the percentage changes in bitcoin prices one to seven weeks later. The fact that they reported *bitcoin* search results looking back four weeks and forward seven weeks should alert us to the possibility that they tried other backward-and-forward combinations that did not work as well. Ditto with the fact that they did not look back four weeks with *bitcoin hack* searches. They evidently tortured the data in their quest for correlations.

Even so, only seven of their fourteen correlations seemed promising for predicting bitcoin prices. Owen Rosebeck and I looked at the predictions made by these correlations during the year following their study and found that they were useless. They might as well have flipped coins to predict bitcoin prices.

Liu and Tsyvinski also calculated the correlations between the number of weekly Twitter *bitcoin* posts and bitcoin returns one to seven weeks later. Unlike the Google trends data, they did not report results for *bitcoin hack* posts. Three of the seven correlations seemed useful, though two were positive and one was negative. With fresh data, none were useful.

The only thing that their data abuse yielded was coincidental statistical correlations. Even though the research was done by an eminent Yale professor and published by the prestigious NBER, the idea that bitcoin prices can be predicted reliably from Google searches and Twitter posts was a fantasy fueled by data torturing.

The irony here is that scientists created statistical tools that were intended to ensure the credibility of scientific research but have had the perverse effect of encouraging researchers to torture data—which makes their research untrustworthy and undermines the credibility of all scientific research.

Data Mining

Traditionally, empirical research begins by specifying a theory and then collecting appropriate data for testing the theory. Many now take the shortcut of looking for patterns in data unencumbered by theory. This is called *data mining* in that researchers rummage through data, not knowing what they will find.

Way back in 2009, Marc Prensky, a writer and speaker with degrees from Yale and Harvard Business School, claimed that

In many cases, scientists no longer have to make educated guesses, construct hypotheses and models, and test them with data-based experiments and examples. Instead, they can mine the complete set of data for patterns that reveal effects, producing scientific conclusions without further experimentation.

We are hard-wired to seek patterns but the data deluge makes the vast majority of patterns waiting to be discovered illusory and useless. Bitcoin is again a good example. Since there is no logical theory (other than greed and market manipulation) that explains fluctuations in bitcoin prices, it is tempting to look for correlations between bitcoin prices and other variables without thinking too hard about whether the correlations make sense. In addition to torturing data, Liu and Tsyvinski mined their data.

They calculated correlations between bitcoin prices and 810 other variables, including such whimsical items as the Canadian dollar–U.S. dollar exchange rate, the price of crude oil, and stock returns in the automobile, book, and beer industries. You might think I am making this up. Sadly, I am not.

They reported finding that bitcoin returns were positively correlated with stock returns in the consumer goods and health care industries and negatively correlated with stock returns in the fabricated products and metal mining industries. These correlations don't make any sense and Liu and Tsyvinski admitted that they had no idea why these data were correlated: "We don't give explanations . . . . We just document this behavior." A skeptic might ask: What is the point of documenting coincidental correlations?

And that is all they found. The Achilles heel of data mining is that large data sets inevitably contain an enormous number of coincidental correlations that are just fool's gold in that they are no more useful than correlations among random numbers. Most fortuitous correlations do not hold up with fresh data, though some, coincidentally, will for a while. One statistical relationship that continued to hold during the period they studied and the year afterward was a negative correlation between bitcoin returns and stock returns in the paperboard-containers-and-boxes industry. This is surely serendipitous—and pointless.

Scientists have assembled enormous databases and created powerful computers and algorithms for analyzing data. The irony is that these resources make it very easy to use data mining to discover chance patterns that are fleeting. Results are reported and then discredited, and we become increasingly skeptical of scientists.

Looking Forward

The three central reasons for the credibility crisis in science are disinformation, data torturing, and data mining. I chose examples related to bitcoin for this Introduction in order to illustrate the crisis. Bitcoin is hardly an isolated case. I chose it because it is so clear that bitcoin believers don't trust authorities and because it so tempting to torture and mine bitcoin data. The chapters to come will expand on these arguments and give examples from a wide variety of fields.

Disinformation

The idea that scientists are a tool used by the ruling class to control us has real consequences. Vaccines for polio, measles, mumps, chickenpox, and other diseases have been lifesavers for millions, yet some people believe that vaccines may be part of a nefarious government plot to harm us or spy on us. Briton Andrew Wakefield is the source of the now debunked claim that the MMR vaccine causes autism. Journals have retracted his research and he has been barred him from practicing medicine in the United Kingdom, yet celebrities continue to spread his wildly irresponsible fabrications throughout the Internet.

So, too, with dangerous falsehoods about COVID-19 vaccines. It is sad enough that people who believe these untruths risk their lives, but they also endanger others. Here are some Internet responses to a March 2021 CNBC news story about COVID-19 vaccinations:

Flu shots are proven to make you 38% more likely to catch another respiratory virus like covid.
SCAMDEMIC
I will do what I want. I will always be free. Dear communist, I'm not sorry
Easy way to target the elderly. Don't be fooled people.
I wont vax i wont mask i wont follow mandates or guidelines and im armed.

Richard Horton, editor of the medical journal *The Lancet*, has noted the paradox between U.S. scientific achievements and the public distrust of scientists:

No other country in the world has the concentration of scientific skill, technical knowledge and productive capacity possessed by the US. It is the world's scientific superpower bar none. And yet this colossus of science utterly failed to bring its expertise successfully to bear on the policy and politics of the nation's response.

Data Torturing

Data torturers have published articles in the *British Medical Journal* reporting that it is dangerous to drive on Friday the 13th, that Asian-Americans are more likely to have heart attacks on the fourth day of the month, and that surgeries are more likely to be fatal if they are done on the surgeon's birthday. Top journals have also published research by respected scientists claiming that pregnant women who eat breakfast cereal daily are more likely to have male babies; people can be cured of deadly diseases by distant healing; and that people do better on tests if they study for the test *after* taking it. The ridiculousness of such conclusions can only undermine respect for science.

Data Mining

Many believe that computer victories over humans in chess, Go, Jeopardy, and other games demonstrate that computers have progressed so far beyond human capabilities that we should rely on their superhuman data-mining prowess to make important decisions for us.

Nope. Computer algorithms are terrific data-mining machines but they are not intelligent in any meaningful sense of the word. Indeed, they do not understand words, including *intelligent*. Computer algorithms are like Nigel Richards, who has won several French-language Scrabble championships without knowing the meaning of the words he spells.

If a data-mining algorithm burrowed through Donald Trump's tweets and found a statistical relationship (and one did) between Trump tweeting the word *with* and the price of tea in China four days later, it would have no way of assessing whether that correlation was meaningful or meaningless. Nor do algorithms have any way of knowing whether it is sensible to hire a person, approve a loan, or predict criminal behavior based on data-mined correlations. Yet they have been used for all of these and their failures erode the credibility of scientific research.

The Irony

These three components of the assault on science were created and fueled by science itself. The disinformation promoted by those who don't trust science is spread far and wide by the Internet that science created. The data torturing that is undermining the integrity of science is the rational response of scientists to the demand for empirical evidence. The temptation to mine data is difficult to resist because it has been made so easy by the mountains of data and powerful computers that scientists created.

It will be hard to fend off these assaults on science, but it is a battle worth fighting.

PART I

Disinformation

CHAPTER 1

The Paranormal Is Normal

Several years ago, while I was writing an op-ed about clinical psychologists, I remembered that my brother, Bob, who has a PhD in psychology, had told me that he once attended a popular presentation at a national convention on how to retain clients for life. That's like lawyers saying that it doesn't matter whether you win or lose a case, but whether the client is happy to pay the bill.

In the middle of these thoughts, my phone rang and it was Bob. I live in California and he lives in Texas, 1400 miles away. We seldom visit each other and only phone on birthdays and holidays and, yet, here he was, calling me while I was thinking about him!

Did my thoughts travel 1400 miles and prompt Bob to call me? That may seem preposterous (to me, it does), but reputable scientists once published an article in a reputable journal reporting that AIDS patients benefitted from prayers sent by distant healers who lived thousands of miles away. To me, that seems preposterous, too.

We have many, many thoughts every day. When I think of people and they don't call, I think nothing of it. Then, one day, Bob calls while I am thinking of him and I remember that phone call for years.

If I believed that thoughts can travel from one person to another and I considered Bob's phone call proof, that would be an example of *confirmation bias*—an inclination to remember events that support our beliefs. I remembered Bob's call for another reason, because it is an example of how a coincidence could be misinterpreted as proof of telepathy—the ability to transmit information from one person to another without using the five physical senses: sight, hearing, taste, smell, and touch.

J. B. Rhine's ESP Experiments

In the 1930s and 1940s, a botanist named J. B. Rhine conducted millions of experiments that were intended to provide scientific support for various paranormal claims, including extrasensory perception (ESP). ESP includes both telepathy (reading another person's mind) and clairvoyance (identifying an unseen object) and can also involve precognition (receiving information about something that has not yet happened).

Zener Card Guessing

Rhine's most well-known experiments involved card-guessing with a "sender" looking at a card while a "receiver" guessed the symbol printed on the card. He began his card-guessing experiments with a standard deck of 52 playing cards but was disappointed with the results and concluded that, instead of guessing the dealt cards, the receivers were reciting their favorite cards.

He switched to a pack of twenty-five cards designed by a Duke psychologist, Karl Zener. The Zener deck consisted of five cards of each of five symbols: circle, cross, wavy lines, square, and star. Using the Zener deck, Rhine reported some truly remarkable results. In his 1935 book, *Extra-Sensory Perception*, Rhine said that 100,000 experiments that he had conducted had "independently established on the basis of this work alone that Extra-Sensory Perception is an actual and demonstrable occurrence."

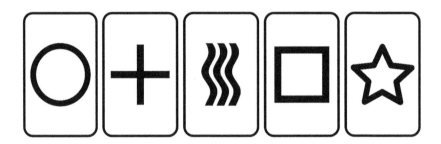

Rhine's book was enthusiastically applauded by eminent reviewers. In two lengthy articles in *Harper's Magazine*, a prominent Columbia University

English professor gushed that "It is hardly too much to say that we may be traveling toward a revolution in the realm of mind more or less comparable to the revolution effected by Copernicus." The co-founder of the Book-of-the-Month Club compared Rhine to Darwin. *The New York Times* science editor wrote that "there can be no doubt of the value of his work ... Dr. Rhine has made a name for himself ... because of his originality, his scrupulous objectivity, his strict adherence to the scientific method."

In 1950, Alan Turing, the celebrated British code breaker and artificial-intelligence pioneer, wrote that

I assume that the reader is familiar with the idea of extra-sensory perception, and the meaning of the four items of it, viz. telepathy, clairvoyance, precognition and psycho-kinesis. These disturbing phenomena seem to deny all our usual scientific ideas. How we should like to discredit them! Unfortunately the statistical evidence, at least for telepathy, is overwhelming.

The public was adapting to some truly incredible scientific inventions at the time and perhaps ESP was another important discovery. Are record players, radios, and telephones less amazing than ESP? Adding to the allure were the many professional mentalists who gave convincing stage performances in which they pretended to have psychic powers.

Rhine capitalized on his growing fame by writing nine books about his experiments and selling packs of "official" Zener cards for people to do their own ESP experiments at home.

Doubts

For scientists, one of the biggest problems with Rhine's reported results was that he was frustratingly vague in exactly how his tests were conducted. As details eventually emerged, so did doubts.

One issue was that there were a number of mathematical errors in Rhine's calculations of the probabilities of various outcomes. He assumed that a guesser would have a 20 percent chance of making a correct guess for each card in a Zener deck. However, the chances of making a correct guess can be improved by taking into account that there are exactly five cards of each design, particularly if the sender identifies the card after each guess. Just as savvy people who play bridge, blackjack, and other card games keep track of cards that have been played, so might receivers keep a mental record of which cards have been selected and which are still in the deck.

A simple card-counting strategy would be to guess the most common symbol left in the deck. If, for example, there are five stars left in the deck and every other symbol has fewer than five remaining cards, guess *star*. When there is only one card left, a card counter is sure to guess correctly. When there are two cards left, a card counter is sure to get both correct if they are identical and has a 50 percent chance of getting both correct if they are different.

Card counting has a big payoff. With Rhine's calculations, the most likely outcome is five correct and the probability of more than eight correct is 0.047. For a card-counter, the most likely outcome is eight correct and the probability of more than eight correct is 0.507.

Another problem is that Rhine did millions of experiments, only some of which were reported. This is an example of what is called the *file drawer problem*, in that successes are reported while failures are tucked in a file drawer and forgotten. Rhine did, literally, have several file cabinets stuffed with the results of experiments he did not report.

Suppose, for example, there is no such thing as ESP but someone correctly guesses the results of ten coin flips. There is only a 1 in 1024 chance of that happening by luck alone but, if we test 1024 people, there is a 63 percent chance that at least one person will be that lucky.

In addition to selectively reporting his own experiments, Rhine reported results that other people sent to him. Besides the lack of controls on these amateur experiments, there is a serious file drawer problem in that those who sent reports of successful experiments to Rhine knew that they might be mentioned in his books but there was less incentive to send Rhine honest reports of unsuccessful experiments.

Rhine's Tenacity

When initial results were disappointing, Rhine and his assistants were absolutely unrelenting in trying to make something out of nothing. For example, they might notice a "U-curve" in which an individual's accuracy rate was above average in the early and later stages of a trial and below average in the middle, or an inverted U-curve in which scores were above average in the middle and below average in the beginning and end.

Sometimes they found intricate patterns of hits and misses and noted that "These complex findings have often been completely unanticipated by the experimenters." The fundamental problem with unanticipated patterns is that *every* large data set has plenty of them. Discovering them only proves that one looked for them.

In addition to looking for correct guesses, Rhine also counted the number of incorrect ones, which he called *psi-missing* or *target avoidance*. One of his justifications for counting misses as evidence of ESP was his conjecture that the receivers knew the cards, but were giving deliberately wrong answers in order to embarrass him.

Rhine also looked for *position effects* (PEs) on an ESP record sheet (Figure 1.1) by dividing the page into halves, quarters, or even finer divisions and then looking for above-chance or below-chance runs in different parts of the record page. He once wrote that

The most common of these PEs are either declines in the scoring rate (across the page from left to right, down the column from top to bottom, or diagonally) or U-curves of success within one of the units of the record page (the set, run, or segment of the run).

Sometimes, Rhine would only report part of an experiment, claiming after the fact that the receiver was not concentrating. In other cases, he would

ESP RECORD SHEET							
Name							
1		2		3		4	
call	card	call	card	call	card	call	card

Figure 1.1 An ESP record sheet.

study the results and discover that, while the guesses did not match the contemporaneous Zener cards, they might have a better-than-chance match with the next card or two cards ahead, as in Figure 1.2. Although none of the guesses in Figure 1.2 match the cards selected at the time of the guess, the first three guesses do match the selections two cards later (*forward displacement*). It was easier for Rhine to believe that some receivers were seeing the future than to accept the inevitability of coincidental patterns.

Rhine also looked for *backward displacement* (guessing cards that had been selected previously). He might find that, even though the overall results did

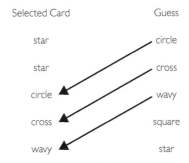

Selected Card   Guess

star   circle

star   cross

circle   wavy

cross   square

wavy   star

Figure 1.2 Forward displacement by two.

not differ from chance, there were correct guesses with forward displacement in some parts of the results and incorrect guesses with backward displacement in other parts. (No, I am not making this up!) His tenacity was impressive; his conclusions were not.

Rhine's dismissal of coincidence as an explanation is tellingly revealed by his recounting of an anecdote similar to my telephone call from my brother Bob:

A prominent federal judge (a Protestant) once dreamed of an elaborate Catholic funeral with many distinctive features, including the exact day of its occurrence. The corpse in the dream was that of President Franklin Roosevelt. But when, 30 days later, the designated day came, the details of the dream were fulfilled by the accidental death of the dreamer's own mother.

And One More Thing

There were other problems. Rhine reported that the best results were obtained in experiments using his patented Zener cards. We might dismiss this as shameless advertising, but he may have been right. It turns out that, when light hit the cards at a certain angle, the design on the front of the card can be seen through the back of the card. Also, in some experiments, the receiver may have seen the face of the cards reflected off the sender's glasses. In other cases, the sender placed the cards face up on the table behind a wooden fence that the receiver may have been able to peek over.

Some of Rhine's most successful card-guessers flopped when observers were present or if they were not allowed to shuffle the cards before the tests, both limitations suggesting that they had cheated during their successful tests. Rhine's explanation for such flops was that ESP is a "weak and delicate" skill that "fades" when skeptics and doubters are near.

Several people who participated in widely publicized successful ESP trials later admitted that they had cheated in various ways. In March 1974 Rhine revealed that he personally knew of at least a dozen incidents of fraudulent tests during the 1940s and 1950s. However, he refused to reveal their names or identify the published papers that reported the bogus results. Rhine said that these incidents were in the past and boasted that "it is impossible for dishonesty to be implemented inside the well-organized psi laboratory today."

Three months later, one of Rhine's star researchers, Walter J. Levy, was forced to resign after he was caught faking experiments. Levy was 26 years old at the time and was Rhine's heir apparent, having just been appointed director of Rhine's Institute for Parapsychology. Levy was conducting astonishing experiments that seemed to demonstrate that gerbils, rats, and unborn chicks could influence a computer's random number generator. Some of Levy's colleagues were skeptical. They secretly connected a second computer to record the results and hid in a closet to observe Levy. They saw him tinkering with the main computer while their hidden computer showed nothing more than random chance. Levy immediately confessed, explaining that he had been under enormous pressure to publish evidence of ESP, and gave Rhine his resignation.

To his credit, Rhine reported Levy's misdeeds and lamented that

When I wrote my paper for the March issue of the Journal, I had not expected to come back to the subject again in publication . . . . Accordingly, I was shocked to discover, only a few months later, a clear example of this same problem, not only right here at the Institute for Parapsychology, but even involving an able and respected colleague and trusted friend.

The most damning indictment of Rhine's experiments is that several researchers redid his experiments and found no evidence of ESP. James Charles Crumbaugh, a psychologist who worked with the Parapsychology Foundation, recounted his frustration:

At the time [1938] of performing the experiments involved I fully expected that they would yield easily all the final answers. I did not imagine that after 28 years I would still be in as much doubt as when I had begun. I repeated a number of the then current [Rhine] techniques, but the results of 3,024 runs [each involving 25 guesses] of the ESP cards—as much work as Rhine reported in his first book—were all negative. In 1940 I utilized further [Rhine] methods with high school students, again with negative findings.

Combaugh issued a challenge in 1959 and reported in 1966 that no one, including Rhine, had accepted it:

> I issued a challenge to all parapsychologists to face up to the repeatability issue. I suggested that they select a single experimental design which offered promise of becoming repeatable and which at the same time permitted full control of all conditions ... To all of this Rhine dissented, denying as always the necessity of satisfying the criterion of repeatability.

Project Alpha

James Randi was a professional magician and ESP skeptic. In 1979 he organized the now-infamous Project Alpha involving the University of Washington's new McDonnell Laboratory for Psychical Research that had been funded by a $500,000 grant from a paranormal believer who was chairman of the board of the McDonnell-Douglas aircraft company. When the McDonnell lab opened, Randi sent the director a memo listing several precautions they should take to avoid being deceived; for example, work with one object at a time and use a permanent marker to ensure that a deceitful person could not switch objects. Randi also offered to travel to the lab at his own expense to observe the experiments, reasoning that a professional magician would know what tricks to look for. The director rejected the offer.

Meanwhile, two teenage magicians (Steve Shaw and Mike Edwards) had contacted Randi and offered their services for testing the lab's experimental controls. Shaw and Edwards independently contacted the lab and over a three-year period became psychic stars as they used several magicians' tricks to fool the researchers into thinking that they had genuine paranormal abilities.

For example, one set of experiments involved the boys pretending to use their mental powers to bend spoons or change the spoons' physical characteristics. The lab had ignored Randi's advice to use a single spoon that was identified with a permanent marker and, instead, used several spoons with paper labels tied to the spoons with strings. After the boys complained that the labels interfered with their mental abilities, the researchers temporarily removed the labels. A little sleight of hand and the labels were switched when they were reattached, giving the impression that the physical characteristics of the spoons had changed. In other cases, a magician would drop a spoon into his lap and bend it with one hand below the table while the researchers were watching the other hand touch a spoon above the table lightly. With another bit of sleight of hand, the spoons were switched and the researchers were stunned.

In another experiment, the magicians were asked to guess the pictures inside sealed envelopes. Incredibly, the envelopes had been sealed with staples and the magicians were left alone with the envelopes. They carefully removed the staples, peeked at the pictures, and put the staples back in the staple holes. In one case, Edwards lost the staples but covered his mistake by opening the envelope himself when the experimenter returned.

In yet another experiment, items were left overnight inside a sealed aquarium in the laboratory. The doors to the laboratory were locked securely with the laboratory director wearing the door keys around his neck. The magicians simply left a window unlocked so that they could break into the lab at night and tinker with the aquarium.

After the magicians revealed their ruse, some observers condemned Randi for unprofessional behavior in trying to deliberately interfere with scientific research—though one might argue that research that was so easily led astray is not really scientific. Some of the researchers remained convinced that the boys' psychic abilities were real and that their announcement of a hoax was the real hoax. The laboratory was subsequently closed, while Shaw later performed as Banachek, a professional magician and mentalist.

Between 1964 and 2015, the James Randi Educational Foundation offered a prize, initially $1000 and raised to $1 million in 1996, to anyone who could demonstrate "under proper observing conditions, evidence of any paranormal, supernatural, or occult power or event." More than a thousand people attempted to win the Randi Prize, but none succeeded.

These issues with paranormal research don't prove that ESP doesn't exist. This is the so-called "black swan" problem. The English used to believe that black swans did not exist because all the swans they had ever seen or read about were white. However, no matter how many white swans we see, these white-swan sightings can never prove that all swans are white. Sure enough, a Dutch explorer found black swans in Australia in 1697.

Similarly, the fact that, up until October 19, 1987, the S&P 500 index of stock prices had never risen or fallen by more than 9 percent in a single day, did not prove that it could not do so. And then it did, falling by 20.3 percent on October 19, 1987, and then, two days later, rising by 9.7 percent.

What we can conclude with some confidence is that much of the evidence in favor of ESP is unconvincing and that, if ESP does exist, it seems to be too rare and weak to be of any practical importance.

Nonetheless, between two-thirds and three-fourths of all Americans believe in some sort of psychic phenomena. A 2011 Economist/YouGov Poll found that 48 percent believe in ESP, while 29 percent do not and 23 percent are unsure.

This disconnect between the views of scientists and the public can only add to the distrust of scientists. As one of my relatives told me, "If scientists don't think ESP is real, then they don't know what they are talking about."

The other takeaway from Rhine's search for evidence of ESP is that some researchers see what they want to see and ignore inconvenient evidence. This is not science.

Who You Gonna Call?

In 1926, before he began his ESP experiments, Rhine and his wife, Louisa Ella Rhine, attended a séance conducted by a famous medium, Mina "Margery" Crandon, who had been praised by Sir Arthur Conan Doyle, the Sherlock Holmes author. Margery claimed that she could channel deceased people and she agreed to put on a demonstration for the Rhines using a "spirit cabinet," which is a large wooden box or curtained-off area that separates the medium from the audience (Figure 1.3).

Figure 1.3 The spirit cabinet illusion.

In the traditional spirit cabinet performance, the medium is tied securely to a chair inside the enclosure with a few props scattered on the floor. When the door is closed or the curtain is pulled, a spirit supposedly visits the medium and the audience hears loud noises like a bell being rung and metal pans clanging together. Some props may be tossed out of the cabinet. Then the cabinet is opened and the medium is still tied to the chair. It seems fairly obvious how the trick is done, but it was a very popular illusion.

I once attended a benefit performance by several magicians and one of the acts was a spirit cabinet illusion performed by two legendary magicians, Frances Willard and her husband, Glenn Falkenstein. After a brief monologue about spiritualism, Falkenstein asked for volunteers from the audience and chose me! Audience volunteers are sometimes stooges who pretend to be innocent spectators but are, in fact, confederates who help the trick succeed. In my case, I really was an innocent spectator.

Willard and Falkenstein did the illusion as a comedy act. Willard's hands were tied with cloth behind her back and she sat on an ordinary chair inside a box formed by four 6-foot-high curtains. The front curtain could be opened or closed, and a bell, tambourine, and several metal pans lay on the floor inside the spirit box. Willard was put into a trance so that she could channel the deceased. Immediately after the front curtain was pulled shut, there was a horrible racket from inside the curtains and several pans were tossed over the curtain and onto the stage. The curtain was pulled open and Willard was seen still in a trance tied to her chair.

I was then instructed to sit on a chair next to Willard so that I could verify that she was tied to her chair the entire time. As soon as I sat down, Willard whispered a request that I play along with the act. I nodded. One of my hands was put on her head and my other hand was placed on her knee. I was then blindfolded and the front curtain was closed, followed again by a loud ruckus and pans flying onto the stage. The curtain was yanked open and Willard was still in a trance tied to her chair, but my pants were now pulled up to my knees and there was a bucket over my head!

Although I was blindfolded, I could feel her hands pulling up my pant legs. I had been asked to cooperate and my hands had been placed on her head and knee so that I would not interfere with her movements (or react instinctively by grabbing her hands as she reached for my pants). A trick knot had obviously allowed Willard to slip her hands in and out of the cloth, yet several people approached me after the show to tell me how much they enjoyed the performance and to ask me how it was done. All I said was that a magician never reveals her secrets.

In the Margery performance that the Rhines witnessed, no one was allowed inside the cabinet with Margery and it was performed with utmost seriousness, with the intention of convincing the audience that a spirit had indeed entered the cabinet and was responsible for what happened. It was clear to the Rhines that Margery was not tightly restrained and that she could easily do everything that happened without the assistance of spirits. The Rhines later wrote an exposé, arguing that "the whole game was a base and brazen trickery, carried out cleverly enough under the guise of spirit manifestations."

The layouts for most purported communications with the deceased are inherently suspicious. Why do mediums hide inside cabinets and why are séances done in the dark where all sorts of trickery is possible? Yet séances and other spiritualism acts used to be wildly popular and taken very seriously. We now know (from confessions by some performers and exposes by Houdini and others) that spiritualism chicanery involved toe tapping, knuckle cracking, table shifting, and all sorts of physical manipulations done under the cover of darkness.

While Rhine dismissed the spirit cabinet as brazen trickery, he later argued that, although mediums were not communicating with the deceased, they were telepathically reading the thoughts of the participants. Those who want to believe can inevitably find evidence supporting their beliefs.

Ghost Seers

Nowadays, the spirit cabinet illusion is performed as a comedy act in the tradition of Willard and Falkenstein. However, there are still plenty of people who believe in ghosts and plenty of people who believe that it is possible to communicate with the deceased.

Paranormal Television

On Halloween evening in 1992, the reliably staid BBC aired a 90-minute show titled *Ghostwatch*. The show appeared to be a live investigation of "the most haunted house in Britain," with BBC presenters Sarah Greene and Craig Charles reporting from the house and Michael Parkinson and Mike Smith (Greene's husband) reporting from a BBC studio.

The show had, in fact, been recorded weeks earlier, but the pretense that it was a live documentary was bolstered by the real BBC presenters, the use of hand-held cameras at the house, and the invitation for viewers to call the BBC's regular phone-in number and talk about their own ghostly experiences.

The (fictional) backstory was that a mother and her two daughters lived in a house that was haunted by a ghost they called Pipes because he banged and rattled the house's old water pipes. The on-scene reporters interviewed the family and neighbors and walked through the house using thermographic cameras, motion detectors, temperature sensors, and other high-tech equipment to look for evidence of poltergeists. They found what they were looking for (and much, much more), but I won't ruin the suspense for those who want to watch the show.

This mockumentary was intended to be a bit of Halloween fun with a few mild scares—ghostly sights and sounds that would create giggles and goose pimples. Nonetheless, the BBC had worried about the possible effects on viewers and they nearly canceled the show at the last minute. Instead, they ran it as part of a BBC drama series called Screen One, with the opening credits showing the name of the person who wrote the script, thinking that this would alert the audience to the fact that this was a fictional drama, not a real documentary. Alas, most viewers thought the show was real.

In the aftermath, one troubled person committed suicide and several children were diagnosed with post-traumatic stress disorder. Although no one was punished, the Broadcasting Standards Commission decreed that "the BBC had a duty to do more than simply hint at the deception it was practising on the audience. In 'Ghostwatch' there was a deliberate attempt to cultivate a sense of menace." Ghostwatch was never again shown on BBC, but the British Film Institute released it as a VHS video and DVD in 2002.

One lasting impact is that Ghostwatch was the inspiration for literally hundreds of paranormal shows that purport to be documentaries, but are, in fact, staged and scripted. There are evidently plenty of viewers with seemingly insatiable appetites for ghost-hunting tales.

Scripted Reality

The appeal of paranormal phenomena is reflected in the growth of paranormal television, sensationalized accounts that are presented as fact. It has even infected programming on the History Channel and Travel Channel, which had previously aired straight-ahead historical and travel documentaries, but now increasingly feature "reality" shows like *Ice Road Truckers*, *Pawn Stars*, and paranormal television programs.

The scripts are formulaic and the effects are cheap but it seems to work. Ironically, part of what makes the shows plausibly realistic is the crude camera work and amateurish acting. It doesn't look or feel like a Hollywood movie. It seems to be ordinary people doing real things.

An astute observer wrote that

reality T.V. is the staple entertainment of the 21st century as it makes us go from viewers to voyeurs . . . We root for the underdogs and the stories that pull at our heartstrings, but we revel in the drama, the fights and the humiliation.

If that sounds like professional wrestling with its heels-and-baby-face drama, it is. Reality television shows, in general, are a lot like professional wrestling, heavily scripted and intended solely to entertain. Even people who know that it is fake enjoy watching with guilty pleasure.

In 2018, a reality show producer revealed that "Ninety-nine percent of the people on reality TV get their expenses covered and maybe a daily stipend of $20 or $30, but that's it." But, hey, fame—no matter how brief—is more important than money. A survey of British teenagers found that nearly one in ten would abandon their education in order to appear on a reality television show.

This producer also confirmed that on purported competition shows, the producers have the final say on who stays on the show and who goes home and that people are chosen for the show to play certain dramatic roles:

I once had a woman cast as a villain who turned out to be the nicest lady ever. As producer, I sat her down and said, "Listen, you were cast in this role. If you want to make good TV, if you want the series to come back and make more money next year, then you need to play along. If you don't, you're going to be cut out entirely." It worked.

For me, one of the big surprises was that they often edit the audio to make it seem that people are saying things that they didn't actually say:

We often take different clips and edit them together to sound like one conversation, sometimes drastically changing the meaning. We can even create complete sentences from scratch. It's so common, we have a name for it: frankenbiting. If you see someone talking and then the camera cuts away to a shot of something else but you still hear their voice, that's likely frankenbiting.

Despite the obvious and not-so-obvious trickery, millions of people are seemingly addicted to trash TV and endlessly amused by fake reality.

On April 1, 2011, *USA Today* reported what might seem to have been an April Fool's joke, but was not. Rutgers University had paid Nicole "Snooki" Polizzi, star of the MTV reality show *Jersey Shore*, $32,000 to give students advice about her GTL (gym, tanning, laundry) lifestyle, which was more than the $30,000 it paid Nobel Laureate Toni Morrison to give the commencement address. Snooki's timeless advice: "Study hard, but party harder."

Ghost shows are a variant that combines the lure of fake-reality shows with a fascination with ghosts. In ghost-hunting shows, viewers are told the legend of a haunted house, hotel, penitentiary, lunatic asylum, or abandoned building and then a camera crew records the ghost hunters spending a night investigating the claims. The light is dim, the searchers are nervous, and the film is grainy. The cast is armed with audio and video recorders and gadgets that measure signs of ghostly activity.

We hear a muffled noise and the cast members tremble. We see a curtain rustle and the cast members stare at each other with frightened faces. We hear a door slammed off camera and some cast members jump and shriek. A blurred face appears in a window. Someone feels a cold breeze. There is a knocking sound as if someone wants to come in or wants the cast to leave. A cast member returns to a room and swears that his hat has been moved from where he left it. There are static-filled noises that seem to be words if you listen closely enough. One person says that the words were "Get out" or "Go away" and it does sound like that when we listen again. Outside, there is often a forest scene with animal sounds that no one has ever heard before. In the morning, the ghost searchers leave frightened and convinced that the ghost rumors are real.

Even though some might be reminded of the Scooby-Doo cartoon show, many take these shows seriously and, after each episode, are a little freaked out and a little more convinced that ghosts are real.

Thousands of people have formed hundreds of ghost-hunting clubs to see for themselves. Some celebrated haunted places charge money for tours. For example, the Trans-Allegheny Lunatic Asylum in West Virginia is now privately owned and offers historical tours, paranormal tours, and overnight ghost hunts. In addition to tours and overnight ghost searches, the Moundsville Penitentiary in West Virginia has an escape room.

Ghost hunters are generally middle class whites with roughly equal numbers of males and females. They dismiss the "scientific" approach of using technology to gather hard evidence, preferring instead to rely on a "sensitive" approach for interpreting thoughts, feelings, and telepathic messages from the deceased.

A 2021 Economist/YouGov Poll found that 41 percent of Americans believe in ghosts while 39 percent do not and 20 percent are unsure. Nearly 75 percent of Americans say they believe in the paranormal. Skeptical scientists say that people who think that they have seen ghosts may have been confused by fatigue, alcohol, drugs, or unusual lighting. Ghost believers say that doubtful scientists are either fools or liars trying to keep the truth from the public.

The Irony

Sometimes, our brain plays tricks on us; sometimes, magicians play tricks on us. More often, imagining the impossible helps us deal with the inexplicable. Bad omens can help us deflect blame. Good omens can help us be more confident. Reading other people's minds can connect us. Communicating with the deceased can help us deal with the deaths of loved ones and can also offer the comforting thought that there is life after death.

The desire to believe can be so strong that the scientific demand for evidence discredits science instead of eradicating superstition. If scientists dismiss common beliefs, believers may well conclude that science is not to be trusted.

CHAPTER 2

Flying Saucers and Space Tourists

Magical things have been seen in the skies for thousands of years, often interpreted by earthlings as religious signs or visitors from other planets. In recent years, unidentified flying object (UFO) sightings have become a popular hobby. Established in 1974, the National UFO Reporting Center has compiled a list of nearly 100,000 UFO sightings, an average of more than five a day. Most sightings have been in the United States, but that may be because the Center is located in the United States. There have been more reported UFO sightings in California than in any other state, indeed more than New York, Texas, and Massachusetts put together. Perhaps aliens prefer California, or maybe California really is la-la land. There have also been more reported UFO sightings in the U.S. on the week of July 4th than during any other week of the year, which shows that aliens like to watch fireworks as much as Americans do. (Joking.)

A Few Too Many Pints

On a foggy January morning in 1966 an Australian farmer named George Pedley heard a hissing sound and when he looked, he saw a flying saucer ascend "in a puff of blue vapor" from a nearby swamp known locally as Horseshoe Lagoon. On closer inspection, he found a 30-foot circle of weeds that had been pressed flat in a clockwise direction as if they had been flattened by spinning air from a flying saucer. The local newspaper ran a story about Pedley's saucer sighting and soon his neighbors discovered five more "saucer nests." These were all near the town of Tully and became known as the Tully nests (Figure 2.1).

Figure 2.1 The original Tully nest.
Credit: UFO Research Queensland.

Local residents told the Royal Australian Air Force that circular weed patterns were common that time of year. An investigation by a local police officer and professors at the University of Queensland concluded that that the circles were most likely caused by short-lived whirlwinds that are called willy-willies in Australia.

Nonetheless, the idea that extraterrestrials were touring Tully resonated with UFO fanatics—and locals were happy to offer lodging and stories to tourists who came to see the Horseshoe Lagoon and squint at faded photographs of saucer nests, much the same way residents of Sedona, Arizona, are happy to advertise New Age energy vortexes:

Sedona has long been regarded as a place both sacred and powerful. It is a cathedral without walls. It is Stonehenge not yet assembled. People travel from all across the globe to experience the mysterious cosmic forces that are said to emanate from the red rocks. They come in search of the vortexes, . . . thought to be swirling centers of energy that are conducive to healing, meditation and self-exploration.

Sedona got its New Age liftoff in 1987, the year of the Harmonic Convergence (a global peace meditation), when 5000 people came to a nearby rock formation called Bell Rock, expecting its top to open and a UFO to emerge. Bell Rock didn't blow its top, but visitors spread the word of Sedona's natural beauty and soon it was a boom town.

A dozen or so years after the Tully nests were reported, designs created by flattened crops of canola, barley, wheat, or other grains began showing up in farmers' fields in southern England. Some suggested that sexually active hedgehogs may have created these crop circles. (No, I am not joking!) Even less plausibly, a molecular biologist named Horace Drew proposed that time travelers from the future were leaving markers to help them navigate our planet (or, at least, England). Never mind that the crop circles disappear when the crops are harvested.

In 1980 a physics professor named Terence Meaden argued that whirlwinds interacting with the local hillsides could produce crop circles, much like the willy-willies in Australia. As the circles grew more complex, so did Meaden's theories—now said to be electromagnetic–hydrodynamic "plasma vortices" created by whirlwinds and enhanced by solar windstorms. A prominent Japanese professor of plasma physics agreed: "Crop circles are a natural phenomenon. They are caused by the plasma vortex." But then crop circles with straight lines began to appear. Explain that with a whirlwind!

Dust devils and bizarre wind patterns might explain simple circles like the Tully nests, but cannot explain the increasingly precise, complex patterns found in England. Aliens and time travelers were popular alternative explanations. Many of the crop circles were in the county of Wiltshire, the home of the Stonehenge and Avebury prehistoric monuments, and a fertile area for tales of haunted houses, demonic dogs, and sacred geometries. Crop-circle enthusiasts ("croppies") began compiling detailed records of crop-circle sightings and traveling thousands of miles to absorb spiritual experiences by hugging wheat stalks or lying down inside a canola circle.

Some people with clearly too much time on their hands claimed to have decoded secret messages, such as "believe," "there is good out there," and "we oppose deception." They did not explain why extraterrestrials would waste their time traveling to Earth to create vapid coded messages. Another purported message was "WE ARE NOT ALONE," which is an odd phrasing for an extraterrestrial message to earthlings. Another theory was that the circles were some sort of secret code related to the pre-Columbian Mayan calendar's Long Count and first Great Cycle, a 5125-year period that would culminate in the end of the world on December 21, 2012. (Spoiler alert: nothing special happened.)

There were doubters and a few obvious questions. Why are crop circles created at night in remote areas that are near roads? Why are aliens so fond of farms in southern England? Why would extraterrestrials travel all the way to Earth simply to create geometric patterns in open fields? Why they would use secret codes if they had something to tell us?

In 1991, two Brits, Doug Bower and Dave Chorley, confessed that they had made hundreds of crop circles using boards with rope handles to smoosh the crops.

One night in 1978 they had been sitting in a pub, drinking warm English beer and talking about the Tully nests when they decided it would be a hoot to create flattened crop circles that made it seem that flying saucers had landed in English fields. Their decades-long prank began and was bought hook, line, and sinker by the croppies.

When they revealed their hoax, Bower and Chorley called a press event and demonstrated how they made crop circles. Afterward, a croppie named Pat Delgado, who had not seen the demonstration, pronounced the circle they had just made to be "authentic" before being told that it was a hoax.

Crop circles have evolved into an elaborate game with pranksters and locals both eager to keep the crop-circle story going. Pranksters like messing with croppies and enjoy creating competitive crop-circle art, enhanced by the thrill of avoiding being caught. There are implicit rules: the circles should be created at night, anonymously, with no incriminating evidence of human involvement. They don't have to be circles but the designs should be interesting. Crop-circle creators now think of themselves as artists and often use sophisticated tools to aid their increasing complicated designs. Photos of the best designs are published in books, catalogues, and calendars.

Farmers are annoyed that parts of their crops are damaged, first by the pranksters and then by trespassing tourists, but many locals are more than happy to sell lodging, tours, and souvenirs to the thousands of New Age visitors who arrive during each crop-circle season. Some croppies make annual pilgrimages in order to experience the spiritual energy of the circles and chat with fellow travelers.

Helicopters, small airplanes, and drones take pictures of the latest crop-circle patterns, which are then uploaded to crop-circle websites. Some towns have pilots flying over their fields daily in order to take pictures of fresh designs that can be downloaded to the Internet and lure more tourists.

Many croppies acknowledge the pranksters but insist that some patterns are too complex to be made by humans working for a few hours in the dark. One croppie reported that "An architect estimated the time needed to create some of the more complex patterns using modern technology. That timeframe was two weeks!" Some rely on spiritual, anti-science arguments:

[T]he paranormal is so loathed by current mainstream science, as this part of it is an affront to our current consensus reality ... [W]e need to be ever-mindful of the fact that it is the non-physical aspects of these subjects that is where the really interesting stuff lies, and we must not be seduced into discarding them in vain hope of "scientific credibility" and "physical proof."

Another argued that "The crop circles are showing us a new science, which we have sadly lost, but ancient cultures practiced."

A regular pilgrim dismissed scientific evidence: "I have my own gauge. I know when they are real because I turn bright red and get flushed. Sometimes I get a little light-headed or feel nauseous. It depends on the frequency, how it interacts with my frequency." And another:

Of course, some are real, and some are hoax, I can tell a real one, by the energy I feel once inside it. Different crop circles seem to resonate different energies, which I usually feel within my chakras .... Each one is different, some seem to energize me, giving me a real buzz, others make me want to lay down in them, and be very still.

Some croppies believe that the pranksters' confessions are the real hoax. Bower now says that he regrets his public confession and wishes that he had allowed the crop circles to maintain their full magical power to amuse and mystify.

The spectacular crop circle shown in Figure 2.2 appeared in July 1996 near Stonehenge. The design is a fractal pattern called the Julia Set and was said to have appeared in less than an hour during the daytime—seemingly ruling out human pranksters. The owner of the field that had been the canvas for this pattern reportedly made about £30,000 charging admission to his field.

Three pranksters later confessed that they had created the pattern in about three hours in the early morning the day before an airplane noticed it. An even more elaborate double-triple Julia Set pattern is shown in Figure 2.3 (pranksters unknown).

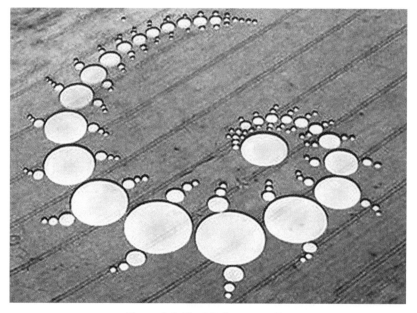

Figure 2.2 The Julia Set crop circle.

Credit: Handy Marks | public domain.

Figure 2.3 A 2001 crop circle in Wiltshire, England containing 409 circles.

Credit: Handy Marks | public domain.

In a bizarre full circle on the crop-circle saga, a 2009 BBC article titled "Stoned Wallabies Make Crop Circles" stated that a government official in the Australian island state of Tasmania had reported that "we have a problem with wallabies entering poppy fields, getting as high as a kite and going around in circles."

Moon-Landing Deniers

The flip side of the belief that aliens from other planets have visited Earth is the conviction that earthling trips to the Moon were a hoax.

In October 1957 the Soviet Union launched a satellite named *Sputnik 1* that sent radio signals for three weeks until its battery died. It continued orbiting the Earth silently for another two months before burning up in the atmosphere. In November the Soviet Union launched *Sputnik 2*, containing a dog named Laika. The dog died on the fourth orbit when the air-conditioning failed, but *Sputnik 2* orbited for another five months before it reentered the Earth's atmosphere and was incinerated.

The space race was on.

The US and the Soviet Union saw space dominance as potentially important for national security and certainly important for bragging rights—a concrete demonstration of the superiority of capitalism or communism. The Soviets sent two more dogs into space while the Americans favored chimpanzees, but the big prize would be awarded for sending a human into orbit and returning safely.

On April 12, 1961, the Soviet Union did just that. Yuri Gagarin orbited the Earth once in *Vostok 1* and returned, ejecting from his spacecraft at 23,000 feet and parachuting into the Soviet Union. There were massive celebrations with April 12 officially declared Cosmonautics Day, which is still observed in Russia and some former Soviet countries.

U.S. President John F. Kennedy had taken office three months earlier and, still smarting from the disastrous Bay of Pigs invasion of Cuba, was determined to win a massive public relations victory. In May 1961, Kennedy delivered a special message to Congress:

[I]f we are to win the battle that is now going on around the world between freedom and tyranny, the dramatic achievements in space which occurred in recent weeks should have made clear to us all, as did the Sputnik in 1957, the impact of this adventure on the minds of men everywhere, who are attempting to make a determination of which road they should take . . .

I believe that this nation should commit itself to achieving the goal, before this decade is out, of landing a man on the Moon and returning him safely to the Earth. No single space project in this period will be more impressive to mankind, or more important for the long-range exploration of space, and none will be so difficult or expensive to accomplish.

On July 16, 1969, six months before the end of the decade, Apollo 11 launched astronauts Neil Armstrong, Buzz Aldrin, and Michael Collins from the appropriately named Kennedy Space Center in Florida. Kennedy's goal was achieved as America won the race to the Moon. A worldwide television audience of a half billion people watched the astronauts land and walk on the Moon and return safely in an ocean splashdown near Hawaii on July 24.

Or did they? Moon-landing conspiracy theorists argue that this initial Moon landing and five subsequent landings involving a total of 12 astronauts were all a giant hoax played on the world by NASA and other government officials.

There had been doubters from the start, but their doubts were bolstered by a 1976 pamphlet titled, *We Never Went to the Moon: America's Thirty Billion Dollar Swindle*. The author, Bill Kaysing, had once worked for Rocketdyne, an aerospace contractor, but only as a technical writer (translating engineering talk into ordinary English). He had a college degree in English but he argued that you didn't need to be a scientist or engineer to know the Moon landing was a hoax.

Before the Apollo 11 launch, he had "a hunch, an intuition, . . . a true conviction" that it would fail. Afterward, he reported that Rocketdyne had estimated the chances of a successful Moon landing and return to be only 0.0017 percent. Kaysing concluded that it would have been easier to fake a Moon landing than to actually do one. He neglected to mention that this estimate had been made in the late 1950s, before a decade of advances made the moon landings feasible.

Kaysing's theory is that NASA launched an empty rocket that crashed after it was out of public view and then faked the return by having a high-altitude army plane drop an empty space capsule into the ocean. The shots of Mission Control in Houston were all faked and the shots purportedly taken on the Moon were filmed on a set in Area 51, the notorious Air Force facility in Nevada that conspiracy theorists believe is the home to UFOs, aliens, and other top-secret projects. The astronauts? They were brainwashed so that they would cooperate and never reveal the hoax.

The Flat Earth Society (there's a name that inspires confidence) claims that the Moon landings were actually filmed at a Disney studio using a script written by Arthur C. Clarke and directed by Stanley Kubrick. Perhaps everyone involved with the film was also brainwashed into 50 years of silence?

Among the most popular evidence cited by the hoax-believers:

1. The astronauts would have died if they had passed through the Van Allen belts of radiation that surround Earth. The reality is that the takeoffs coincided with the belts' lowest intensity and the spacecraft were specially insulated to protect the astronauts. Their radiation exposure was comparable to a chest X-ray.

2. The Moon walks must have been filmed in a studio because there are no stars in the Apollo photographs. The reality is that the photographs were taken during lunar daytime. Similarly, photographs taken on Earth during the daytime do not show stars. On other occasions space-shuttle astronauts have taken long-exposure nighttime photos that show plenty of stars.

3. There is no air on the Moon, yet the American flag planted on the Moon seems to be fluttering in the wind. The reality is that, knowing that there is no air on the Moon, NASA built the flag post with the top of the flag attached to a horizontal rod to hold the flag in an open position. When the astronauts planted the flag post in the ground, they twisted it back and forth to dig into the lunar soil—which left the flag in a ripple shape. Videos of the flag after it had been planted clearly show that it is not blowing in the wind.

One very big counter-argument to the Moon-landing deniers is the implausibility of every single one of the tens of thousands of people involved keeping a hoax a secret for more than fifty years. Another counter-argument is that the astronauts brought back solid proof of their landing: Moon rocks. As a goodwill gesture, samples have been given to 135 countries, many of whom tested the rocks and confirmed that they are not from Earth.

For example, several Moon rocks contained glass spherules (Figure 2.4) that can be created by meteoric impacts that melt and vaporize rocks. Unlike those found on Earth, glass spherules on the Moon are in pristine condition because there is no atmosphere or water to cause weathering.

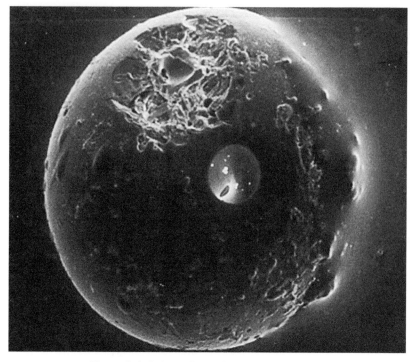

Figure 2.4 A glass spherule from the Earth's Moon.
Credit: NASA

The hoax theory got a big boost in 2001 when Fox TV twice aired a prime-time special report titled *Conspiracy Theory: Did We Land on the Moon?*, hosted by Mitch Pileggi, who was an actor on *The X-Files*, a Fox television series about a government conspiracy to hide the truth about UFOs and extraterrestrials. There is an initial half-hearted disclaimer:

The following program deals with a controversial subject.
The theories expressed are not the only possible interpretation.
Viewers are invited to make a judgment based on all available information.

The show itself presents the claims of Kaysing and other "experts" and unconvincing NASA denials. Unfortunately, this completely misleading show can still be viewed on streaming services, and is labeled a documentary even though it is no more documentary than are ghost-hunting shows.

An interesting footnote is that one of the Fox show experts (Bart Sibrel) reportedly harassed some of the astronauts publicly. In 2002 he lured Buzz Aldrin to a Beverly Hills hotel under the false pretense that he was going to be interviewed by a Japanese children's television show. When Aldrin showed up, he was confronted by a camera crew and Sibrel waving a Bible in Aldrin's face, demanding that he swear on it that he walked on the Moon. Aldrin tried to evade him but Sibrel persisted, calling Aldrin "a coward, and a liar, and a thief," at which point 72-year-old Aldrin punched 37-year-old, 250-pound Sibrel in the face. Sibrel immediately turned to his camera crew and asked, "Did you get that?," evidently hoping that Aldrin would be in legal trouble. The district attorney's office did not pursue charges because Aldrin had a clean record, Sibrel sustained no visible injury, and witnesses testified that Sibrel had been the instigator.

When Neil Armstrong stepped onto the Moon's surface, he famously said, "That's one small step for a man, one giant leap for mankind." A news story about Aldrin's confrontation with Sibrel had the subheading "One small punch for a man, one giant blow to conspiracy theorists."

In 2018, a Brit named Martin Kenny argued that the Moon landings were obviously faked because the Moon is made of light, rendering a Moon landing impossible. He concluded, "In the past, you saw the moon landings and there was no way to check any of it. Now, in the age of technology, a lot of young people are now investigating for themselves." Exactly how are young people going to check whether a Moon landing done fifty years ago was real or fake? By surfing the Internet?

The Internet is the problem, not the solution. A former chief historian of NASA noted that "The reality is, the internet has made it possible for people to say whatever the hell they like to a broader number of people than ever before."

A 2019 YouGov poll of British adults found that 4 percent definitely agreed with the statement that "The moon landing was staged," another 16 percent thought the statement was probably true, and 9 percent said they didn't know one way or another. Similar polls in the United States and Russia have given similar results, with 20 to 30 percent thinking the landings were probably or definitely faked.

Turning Kaysing's argument on its head, it is evidently easier for some people to believe that scientists are capable of an elaborate hoax than to believe that scientists are capable of a real accomplishment.

Our addiction to Internet videos and our affinity for conspiracy theories don't help. Do an Internet search for "moon landing," "moon walk," or "Apollo" and soon you will be inundated with Moon-hoax "documentaries." Watch one and your computer will suggest a dozen more.

YouTube celebrity Shane Dawson has aired several conspiracy videos that target teens and children, including "Moon Landing Conspiracy Theory." The video is a goofy, painfully amateurish rehash of the "evidence" cited by Moon-landing deniers. Dawson concludes: "It's not a shock, because the government fakes so much shit. I mean, we've talked about 9/11, we've talked about crisis actors. Why wouldn't the moon landing be fake?" For good measure, he adds, "What if the moon is a hologram? What if it is not real?" Yet, this nonsense has had more than 7 millions views with nearly 300,000 likes and fewer than 5,000 dislikes.

The Irony

Scientists didn't create the Internet and social media in order to discredit science, but that is what is happening. UFOs and faked Moon landings are two sides of the same coin—the belief that governments and scientists are lying to us and cannot be trusted. The supposed proof of these lies is on the Internet for the distrustful to see and to share via social media—including photos and videos fabricated by incredible editing tools that were developed by scientists but can be used by anyone to distort reality.

CHAPTER 3

Elite Conspiracies

In a 2016 video-streamed Q&A town hall event, Facebook CEO Mark Zuckerberg was asked, "Mark, are the allegations true that you're secretly a lizard?" Zuckerberg answered, "I'm gonna have to go with 'no' on that," though he did lick his lips as he proclaimed, "I am not a lizard."

Zuckerberg is hardly the only public figure to be accused of being reptilian. In 2014, a New Zealander filed an Official Information Act request for Prime Minister John Key to provide "any evidence to disprove the theory that Mr. John Key is in fact a David Icke style shape shifting reptilian alien ushering humanity towards enslavement." Icke (now, there's a reptilian name) is a professional conspiracy theorist who claims that many world leaders (including (now deceased) Queen Elizabeth, George W. Bush, the Clintons, and the Illuminati) are shape-shifting reptiles from the constellation Draco.

During a television interview, Prime Minister Key dutifully declared that

To the best of my knowledge, no. Having been asked that question directly, I've taken the unusual step of not only seeing a doctor but a vet, and both have confirmed I'm not a reptile . . . [I'm] just an ordinary Kiwi bloke.

In 2020, Icke was permanently banned from YouTube and Facebook after posting videos claiming that COVID-19 was being spread by 5G technology and that COVID-19 vaccines would implant "nanotechnology microchips" in humans to control them.

As much as one might suspect that Icke is a publicity-seeking fraud, his wildest claims resonate with some people. A 2016 Public Policy Survey found that 4 percent of Americans (12.5 million people) believe that the U.S. government is controlled by alien lizard people. Millions believe that 5G is being used to spread COVID-19 (and they have burned down cell towers to stop it) and that COVID vaccines are a nefarious plot (and they refuse to be vaccinated). A recent survey found that 44 percent of Republicans, 24 percent of independents, and 19 percent of Democrats believe that Bill Gates is developing a COVID-19 vaccine that will implant microchips in us so that our movements can be monitored.

Our lives are filled with unforeseen events. Rather than accept the fact that we are buffeted by unpredictable occurrences, we try to impose order by attributing our misfortune to others—most often, to elites who are thought to conspire to manipulate and control us and enrich themselves at our expense.

The Illuminati

The advice column in the April 1969 issue of *Playboy* included the usual questions about broads, beers, and baldness, followed by an unusually long (300-word) letter that began:

I recently heard an old man of right-wing views—a friend of my grandparents—assert that the current wave of assassinations in America is the work of a secret society called the Illuminati. He said that the Illuminati have existed throughout history, own the international banking cartels, have all been 32nd-degree Masons and were known to Ian Fleming, who portrayed them as SPECTRE in his James Bond books—for which the Illuminati did away with Mr. Fleming.

The letter ended with two questions:

Do they really own all the banks and TV stations? And who have they killed lately?

Playboy gave a 350-word response tracing the history of the Illuminati and seemingly reassuring the letter-writer:

The belief that the Illuminati . . . are responsible for most of our evils is about the fourth most common form of organized paranoia extant (its three more popular rivals are the Elders of Zion conspiracy, the Jesuit conspiracy, and the notion that we have already been invaded by outer space, our governments being in the hands of Martians).

This exchange was followed in later issues by more letters and answers arguing for and against previous letters and answers.

Why was so much valuable space devoted to such a barely relevant topic for *Playboy* readers? Did the editors run out of sexy photographs?

It turns out that the letters were an elaborate prank played by Robert Anton Wilson, who was a *Playboy* editor at the time, and Kerry Thornley (aka Lord Omar Khayyam Ravenhurst), one of the authors of the mischievous book *Principia Discordia*, which urges readers to worship Eris, the Greek goddess of chaos, through civil disobedience and pranks. Wilson ridiculed dogma and often said, "belief is the death of intelligence."

As for the Illuminati, the Order of the Illuminati was a secret society founded in Bavaria, Germany, in 1776 by intellectuals opposed to the power of the Bavarian monarch and the Roman Catholic Church. Within a decade, it had been infiltrated and disbanded by the Bavarian government.

Fast forward to the 1960s and the merry pranksters, Wilson and Thornley, decided it would be fun to get people riled up about this defunct secret organization. They wrote fictitious letters to *Playboy*, initially inquiring and then arguing about the Illuminati, which Wilson ensured would be published. *Playboy* was an unlikely venue, but that is where Wilson worked.

One thing led to another and soon Wilson and his good friend, Robert Shea, who was also a *Playboy* editor, published *The Illuminatus! Trilogy*, a free-for-all collection of fanciful conspiracy theories, including the claim that George Washington was actually a reincarnation of Illumanati founder Adam Weishaupt. They later wrote that their working assumption was that "all these nuts are right, and every single conspiracy they complain about really exists."

This book of nonsense was a surprise success and soon there was an Illuminati play, an Illuminati card game, and plenty of conspiracies to attribute to the Illuminati, including the assassination of JFK and the 9/11 terrorist attacks. The imaginary Illuminati continue to make guest appearances in plays, movies, novels, and comic books, oddly enough with their purported goal no longer to oppose the ruling elite but to impose a new world order. A 2019 survey found that 21 percent believe "that the Illuminati secretly control the world."

When Joe Biden was sworn in as the 46th President of the United States, social media lit up with claims that he used a Masonic/Illuminati Bible. One widely shared Facebook post mused,

Sooo has anyone else realized this yet or???? Masonic/Illuminati Bible that Biden swore on yesterday . . . I do believe a lot of weird and scary shit is gonna go down at some point.

No matter that Biden was holding a 1893 edition of the Douay–Rheims Bible which, until the 1900s, was the only authorized English-language Bible for Roman Catholics. This Bible was a family heirloom that Biden had used at nine previous swearing-in ceremonies (seven times as a U.S. Senator and twice at

Vice President). John F. Kennedy, the only previous Catholic President, had also used a Douay–Rheims Bible when he took his oath of office in 1961. The idea that the Douay–Rheims Bible is an Illuminati Bible is preposterous since the Illuminati opposed the Roman Catholic Church, but logic doesn't matter to people who only see what they want to see.

Almost immediately after I wrote an op-ed about Illuminati conspiracy theories, I began receiving e-mails from conspiracy fanatics. One had the subject line "so what if we rename the Illuminati the Deep State?" and listed thirteen things I should investigate in order to educate myself about conspiracies. Most were related to the claim that Donald Trump's "landslide victory" in the 2020 presidential election was stolen from him:

> How come the head of the accounting system at Dominion belong to Antifa?
>
> How did a board member of George Soros come to control the software used to count votes in swing states?
>
> How did data from Dominion systems flow back and forth to IP addresses in Iran, Russia and the vast majority with China?

I was torn between: (a) feeling sad that my correspondent spent so much time looking for conspiracy evidence; and (b) acknowledging that he had a hobby that he enjoyed and evidently kept him busy. I was certain that, like avoiding the eyes of crazed druggies, I was not going to encourage him by replying to his e-mail.

Not counting pranksters, what is the attraction of conspiracy theories? A series of surveys between the years 2006 and 2011 found that roughly half of Americans believe in at least one conspiracy theory. In the 2011 survey, for example, 19 percent believed that

> Certain U.S. government officials planned the attacks of September 11, 2001, because they wanted the United States to go to war in the Middle East

and 25 percent believed that

> The current financial crisis was secretly orchestrated by a small group of Wall Street bankers to extend the power of the Federal Reserve and further their control of the world's economy.

A New World Order

The Illuminati conspiracy theory is but one of many variations of the widespread belief that a group of elites have taken (or are in the process of taking) control of businesses, banks, governments, and the media in order to create a New World Order in which the ruthless elite use powerful computers to control people everywhere. Many disparate things are said to be part of the plot: George Soros, the Clintons, Federal Reserve, Trilateral Commission, Council on Foreign Relations, the Protocols of the Elders of Zion, Bohemian Grove, Bilderberg Group, Skull & Bones, the Denver Airport, and the list goes on and on.

The stories are outlandish and entertaining, yet they evidently resonate with many—perhaps because people struggling with personal difficulties find it easier to blame their troubles on others than to take responsibility. For conspiracy believers, bad things happen because bad people make them happen. If I buy a stock and the price goes down, it is because insiders are manipulating the market. If I lose my job, it is because the Fed is manipulating the economy. If my life isn't working out the way I wanted, it is because the elite control the government for their own benefit.

Vaccines for polio, chickenpox, measles, mumps, and other childhood diseases have been so effective that, until recently, only the elderly remembered that these contagious diseases were deadly, yet some people believe that vaccines may be part of a government conspiracy to harm us or spy on us. People who believe these untruths are not only risking their health, they are also endangering others. Ironically, science discovered the vaccines and also created the Internet that allows conspiracy theories to spread like an unstoppable virus.

The Wealthy Jewish Puppeteer

In theory, people won't pay $3 for a can of chicken soup if they can buy the same soup at a nearby store for $2. Applied internationally, identical (or nearly identical) goods should cost about the same anywhere in the world. Suppose, for example, that the exchange rate between British pounds and U.S. dollars is 1.50 dollars/pound. If a sweater sells for £20 in Britain, essentially identical sweaters should sell for $30 in the United States. If the U.S. price were higher than $30, Americans would import British sweaters instead of buying overpriced American sweaters. If the U.S. price was less than $30, the English would import American sweaters. This is called the *law of one price*.

The law of one price works best for products that are relatively homogeneous and can be imported relatively inexpensively. The law of one price doesn't work very well when consumers think the products are different (for example, BMWs and Chevys) or when there are high transportation costs, taxes, and other trade barriers. Wine can be relatively expensive in France if the French tax wine imports heavily. A haircut and round of golf in Japan can cost more than in the United States because it is impractical for the Japanese to have their hair cut in Iowa and play golf in Georgia.

When the law of one price is applied to entire countries, it predicts that exchange rates depend on relative rates of inflation. If prices increase by 5 percent in the United Kingdom and only 3 percent in the United States, British goods will become more expensive unless the value of the pound falls by 2 percent relative to the dollar. More generally, countries with relatively high rates of inflation should experience depreciating currencies.

This theory is used by some currency traders to predict movements in exchange rates. If there is 5 percent inflation in England and 3 percent inflation in the United States but the dollar/pound exchange rate doesn't move, speculators might use pounds to buy dollars with the expectation of selling these dollars after the value of the dollar increases.

Currency speculation is a zero-sum game. If a speculator uses pounds to buy dollars (expecting the dollar to appreciate) from another speculator (who expects the pound to appreciate), any profit that either makes will be exactly offset by losses for the other speculator.

A dramatic example occurred in the 1990s, when George Soros made a huge bet on the British pound relative to the German Deutschmark. Soros was named György Schwartz when he was born in Budapest in 1930 to Jewish parents whom Soros later described as essentially antisemitic. As Hungary itself became increasingly antisemitic, the family changed their name from Schwartz to Soros. After the War, he moved to England and earned Bachelor's and Master's degrees in philosophy from the London School of Economics.

After a variety of odd jobs, Soros was able to get a toehold in the financial industry and eventually establish several successful investment funds. What made his reputation—for better or worse—was his 1992 speculation on the British pound.

Eight European countries (Belgium, Denmark, France, Germany, Ireland, Italy, Luxembourg, and Netherlands) formed the European Monetary System in 1979 with an Exchange Rate Mechanism (ERM) agreement to stabilize their exchange rates relative to Germany's Deutschmark at specified target rates, plus or minus 6 percent. If an exchange rate threatened to move outside that range, the member countries would buy or sell currencies as needed to keep the rate inside the range.

Britain joined on October 8, 1990, with a target exchange rate of 2.95 Deutschmarks to the British pound. Figure 3.1 shows that its exchange rate had fluctuated around the value 2.95 prior to Britain's entry in October 1990. However, the rate of inflation in Britain had averaged 6.7 percent over the previous three years compared to 2.6 percent in Germany and many currency traders, including Soros, concluded that a depreciation of the pound relative to the mark was inevitable.

Figure 3.1 The deutschmark/pound exchange rate, 1988–95.

A junior economist in Britain's Treasury department later wrote that "the fundamental problem was that we'd joined the ERM at the wrong rate; sterling was overvalued, meaning that we were stuck with a structural [trade] deficit" because the overvalued pound made British goods too expensive to compete with foreign goods. What set Soros apart was that he was willing to go all-in on his belief that the pound would depreciate. As he later wrote, "There is no point in being confident and having a small position."

Soros borrowed $10 billion worth of British pounds which he used to buy German marks at an exchange rate of around 2.95. (He also placed smaller bets on the French franc, British stocks, and German and French bonds.) If the pound fell 10 percent relative to the mark, he could use his marks to buy $11 billion worth of pounds and make a $1 billion profit, less the interest on his loan. Since his loan was backed by around $1 billion in collateral, he would essentially double his money if the pound fell 10 percent. On the other hand, he would have been bankrupted if the pound appreciated 10 percent. As Keynes wryly observed, "markets can remain irrational longer than you can remain solvent."

Soros later wrote that, "Markets can influence the events that they anticipate." As Soros and like-minded currency speculators bought marks with their
borrowed pounds, they were pressuring the mark to appreciate relative to the
pound. To prevent the depreciation of the pound, the Bank of England and
other sympathetic European central banks were forced to buy pounds. Since
currency speculation is a zero-sum game, any profits the speculators made from
selling pounds would be matched by losses for the central banks that were
buying pounds. The British Treasury reportedly bought $27 billion worth of
pounds and would suffer a $2.7 billion loss if the value of the pound fell 10
percent.

On the evening of September 16, 1992 ("Black Wednesday"), the British finance minister announced that Britain would abandon the ERM and let the
pound's exchange rate be determined by market forces. The market price soon
settled at around 2.50 DM/£, some 15 percent below the 2.95 target.

Soros' funds made a $7 billion profit from their various wagers and Soros
himself made nearly $1.5 billion. He became known forever as "the man who
broke the Bank of England."

Soros was later rumored to have been involved in currency speculation with
Finnish markkas and various Asian currencies. Some of his giant wagers have
been giant losses. In 1994, he lost hundreds of millions of dollars betting that
the Japanese yen would depreciate relative to the U.S. dollar. He also reportedly
lost a billion dollars on investments that he made based on the assumption
that Donald Trump would lose the 2016 U.S. presidential election and, then,
when Trump did win, betting that the U.S. stock market would collapse. Of
course, losing $1 billion when you have $30 billion is less consequential than
making $1 billion when you only have $1 billion. As I write this, Soros is
92 years old with an estimated net worth of $8.6 billion, after having given away
$32 billion.

There is something unsavory about profiting from financial crises that you
create. In 1999 Paul Krugman wrote that "There really are investors who not
only move money in anticipation of a currency crisis, but actually do their best
to trigger that crisis for fun and profit. These new actors on the scene do not
yet have a standard name; my proposed term is 'Soroi'" (his made-up plural of
Soros).

More sinister is the perception fueled by conspiracy devotees that the billions of dollars that Soros has given away are not for humanitarian causes, but
part of a nefarious master plan to control world businesses and governments.
In 2010, Fox News ran a three-part series on "the puppet master, George Soros"

in which he is portrayed as the archetypical Jewish financier who controls politicians around the globe.

A 2011 survey found that 19 percent of Americans believed that "Billionaire George Soros is behind a hidden plot to destabilize the American government, take control of the media, and put the world under his control."

Big Brother is Watching

The Vietnam War dragged on for nearly 20 years, from November 1955 until April 1975, with North Vietnam (supported by the Soviet Union, China, and other communist countries) battling South Vietnam (supported by the United States and other anti-communist countries). In 1967, Robert McNamara, the U.S. Secretary of Defense, created a special task force to write a history of the Vietnam War up until that point. The 3000-page study (with 4000 pages of supporting documents) was completed in January 1969, shortly before president-elect Richard Nixon took office. The report was classified "Top Secret" and only 15 copies were made.

The report provided a candid assessment of the war, including a recounting of several ways in which the U.S. government had misled the American people. For example, the U.S. had secretly aided France's battles against Vietnamese revolutionaries from 1948 to 1954 and had itself waged covert battles against the communist Vietnamese prior to the official declaration of war.

The report also noted that, during the Kennedy administration, the U.S. had been intimately involved in the 1963 military coup that assassinated South Vietnamese president Ngô Đình Diệm:

Beginning in August 1963 we variously authorized, sanctioned and encouraged the coup efforts of the Vietnamese generals and offered full support for a successor government.

While President Lyndon Johnson was proclaiming the war's goal to be the defense of an "independent, non-Communist South Vietnam," McNamara was privately telling Johnson that the war was part of the government's long-term policy to contain China:

China—like Germany in 1917, like Germany in the West and Japan in the East in the late 30s, and like the USSR in 1947—looms as a major power threatening to undercut our importance and effectiveness in the world and, more remotely but more menacingly, to organize all of Asia against us.

Daniel Ellsberg, an economist and military analyst working for RAND, obtained a copy of the report and made photocopies that he passed on to *The New York Times, Washington Post*, and other newspapers, hoping that their publication would help end "a wrongful war."

In April 1971, *The New York Times* began publishing a series of front-page articles based on these photocopies, which came to be known as the Pentagon Papers. The contents were explosive because these were official government documents revealing that the U.S. government had systematically lied to its citizens about the origins of the increasingly unpopular Vietnam War.

President Nixon's initial reaction was perhaps satisfaction at the damage done to the reputations of Kennedy and Johnson, whom he disliked intensely. However, Nixon advisor Henry Kissinger persuaded him that it would set a bad precedent if top-secret government documents were allowed to be made public. The government charged Ellsberg with espionage, theft, and conspiracy and obtained a federal court injunction ordering *The New York Times* to cease publication of the Pentagon Papers.

The charges against Ellsberg were eventually dismissed because of government misconduct. In an odd foreshadowing, E. Howard Hunt and G. Gordon Liddy (nicknamed "the plumbers," because they were tasked with fixing leaks) broke into the offices of Ellsberg's psychiatrist in an unsuccessful attempt to discover embarrassing information that would discredit Ellsberg. The following year, they were involved in the Watergate break-in that led to Nixon's resignation.

The *New York Times* and *Washington Post* appealed the cease-publication injunction and the case quickly moved to the Supreme Court with the Court ruling 6–3 for the newspapers. In a particularly passionate opinion, Justice Hugo Black wrote that

[T]he injunction against *The New York Times* should have been vacated without oral argument when the cases were first presented . . . . [E]very moment's continuance of the injunctions . . . amounts to a flagrant, indefensible, and continuing violation of the First Amendment. . . . The press was to serve the governed, not the governors.

Freedom of the press is enshrined in the First Amendment because of the belief that a nation's citizens should know what their government is doing for them and to them, and the hope that governments will serve their citizens better if they know that they are being held accountable.

Technology has the power to enhance the freedom of the press. Today's Daniel Ellsbergs do not need to lug around 7000-page documents or rely on newspapers to publicize evidence of government misdeeds. They need only have a digital copy on a flash drive or in the cloud and they need only post PDFs on the Internet and use social media to announce their existence.

Unfortunately, technology also has the power to allow repressive governments to spy on their citizens and suppress dissent more effectively. Too often, technology is being used to serve the governors, not the governed.

The Chinese government monitors virtually everything its citizens do on the Internet and on their phones and uses this information to calculate "social credit" scores that reward obedient citizens with discounts and privileges and punish those who are perceived to be untrustworthy by restricting what they can buy, where they can live, and where they can travel. In Xinjiang, where half the population is Muslim Turkic peoples, this electronic monitoring is supplemented by thousands of CCTV cameras, and millions who have been deemed to have low "reliability status" have been killed or imprisoned in re-education camps.

China is hardly alone, nor is incarceration needed to suppress dissent. People will self-censor if they know (or even suspect) that they are being monitored and may be punished.

Edward Snowden was a computer intelligence consultant working in Hawaii for the National Security Agency (NSA) in 2013 when he saw James Clapper, Director of National Intelligence, respond to a question from U.S. Senator Ron Wyden about whether NSA collects "any type of data at all on millions or hundreds of millions of Americans." Clapper answered, under oath, "No sir, not wittingly." Snowden knew that NSA was in fact using Facebook, Google, and Microsoft servers to track online activity and forcing U.S. telecommunications companies to give NSA information on virtually every U.S. phone call and text message, and was also sharing global surveillance information with Australia, Canada, the United Kingdom, and New Zealand as part of the Five Eyes Program.

Three days later, Snowden flew to Hong Kong and gave copies of thousands of classified NSA documents to several journalists who then reported stories of government snooping in the *Washington Post*, *Guardian*, and other publications. Like Ellsberg, the U.S. government charged Snowden with espionage and conspiracy.

Fearing extradition from Hong Kong, Snowden tried to fly to Latin America, with stopovers in Moscow and Havana. Soon after his fight took off, the U.S. government revoked Snowden's passport and he ended up stranded in Moscow. After nearly a decade in Russia, he was granted Russian citizenship in 2022.

On September 2, 2020, a U.S. federal court of appeals ruled unanimously that the bulk collection of telephone data that had been revealed by Snowden violated the Foreign Intelligence Surveillance Act and potentially violated the Fourth Amendment's prohibition of unreasonable searches and seizures. NSA's bulk data collection program has reportedly ended, though you never know—and that's the point. If people believe that the government may be monitoring every word they say or write, they will self-censor and freedom of speech will effectively be silenced.

After Snowden revealed the NSA's online monitoring, searches for keywords like "jihad" and "chemical weapon" abruptly declined—no doubt because people feared that they might arouse the suspicions of the NSA and have all of their activities monitored more closely.

One of the most popular storylines in the widely acclaimed television show *The Good Wife* is NSA techies being entertained as they eavesdrop on the heroine's personal life. It clearly resonated with viewers and reinforced their fears that the NSA might be listening to their own conversations.

Instead of bolstering freedom of the press by fostering meaningful political discussions, the Internet is too often used by governments to monitor citizens more closely and effectively squash free speech. Instead of disseminating real news and debating important ideas, the Internet is too often used to promote whacky theories and titillate users with inane gossip.

Animal Spies

A variety of animals have been used in military operations. One advantage is that people may not expect a bird to be carrying a hidden camera or a cat to have an implanted listening device. Another advantage is that animals can go places humans cannot; for example, insects crawling through tight spaces or whales making deep ocean dives. In addition, the loss of a locust or vulture may be less tragic than the loss of a human life.

Many surveillance programs are top secret, but we know about a few; for example, the U.S. and Russia use seals, beluga whales, sea lions, bottlenose dolphins, and other marine mammals to look for underwater mines and swimmers in restricted areas. In 2011 the U.S. Navy did a media demonstration where a retired (human) SEAL made five attempts to enter a San Diego harbor patrolled by dolphins and sea lions. He was caught every time and, "The sea lion even managed to attach a clamp to the diver's leg, and handlers on the surface reeled him in like a fish."

Legend has it that during the Napoleonic Wars, a storm caused a French ship to be wrecked off the coast of Hartlepool, a town in northeast England. The only survivor was a monkey dressed in a French military uniform (evidently to entertain the French troops). The local English bumpkins, having never seen a monkey or a Frenchman, assumed that the monkey was a French spy and hung it from the ship's mask on the beach.

Hartlepoolians have subsequently been mocked as "monkey hangers," but the local soccer club and two rugby teams celebrate their monkey-hanging history. The Hartlepool Rovers rugby team has a crest featuring a hung monkey. The soccer team has the nickname "monkey hangers" and their mascot, a person wearing a costume of a monkey in a soccer uniform, is named "H'Angus the Monkey." Stuart Drummond, one of the costumed mascots, campaigned for mayor wearing his costume as a publicity stunt in 2002; his only campaign promise was free bananas for all school children. He won.

Credit: Hartlepool Rovers.

In the Middle East, pigeons, vultures, eagles, storks, and squirrels have been arrested or shot by various government authorities for spying. Egyptian authorities once accused Israel of using sharks to attack swimmers in order to hurt the Egyptian tourism industry. Palestinian authorities accused Israel of unleashing wild boars to destroy crops, and a Jordanian television host accused Israel of unleashing Norwegian rats carrying the Bubonic plague. During Russia's 2022 invasion of Ukraine, Vladimir Putin accused Ukraine of using migratory birds and insects to transmit biological weapons into Russia.

The U.S. military is, in fact, known to be working on cyborg insects for smelling explosives and carrying miniature cameras and listening devices but when it comes to paranoia, fact is often no match for fiction.

"Birds Aren't Real! Pass it on! . . ."

In 2017, a college student started the "Birds Aren't Real" conspiracy theory. The claim is that real birds were exterminated by the U.S. government and replaced with drones that are disguised as birds and monitor us from the sky—which gives "bird watching" a whole new meaning.

Some facts underlie the conspiracy claims. For more than a century, pigeons have been used to take aerial photographs with cameras strapped to their bodies (Figure 3.2). Surely, the theory runs, the government can now make robotic birds with cameras hidden inside.

Figure 3.2 A pigeon photographer.
Credit: Deutsches Spionagemuseum Berlin/German Spy Museum Berlin.

"Birds Aren't Real" started as a joke and morphed into a marketing opportunity for T-shirts and hats. Now, some people actually believe it. Even if you recognize "Birds Aren't Real" as satire, the fact that you thought about it for more than a second says something about your propensity for paranoia.

24/7 Surveillance

Since at least 2013 governments around the world–including the United States, India, the U.A.E., and almost all European governments–have reportedly been licensing Pegasus smartphone spyware from an Israeli company named NSO Group. Once it has infected a phone (and it is surprisingly easy to do so; in many cases it is "zero click," not dependent on the phone user doing anything), Pegasus can monitor virtually everything done on the phone, including copying all text messages and photos and recording all phone calls. It can even activate the phone's microphone and camera to record and film the phone's owner. Some governments have no doubt developed similar spyware.

NSO claims that the governments that license Pegasus are contractually prohibited from using the spyware for anything other than combatting "serious crime and terrorism;" however, Pegasus is being used in several countries to spy on dissidents and is even used by drug cartels to spy on journalists and police officers. In 2021 the *Guardian* reported that a leaked list of Pegasus targets included more than 180 editors, journalists, and reporters working for the *Financial Times, New York Times, Wall Street Journal,* the *Economist,* CNN, Bloomberg, Reuters, and other publications. Pegasus has reportedly infected the phones of hundreds of politicians, including the office of Britain's Prime Minister.

NSO is, of course, not the only peddler of sophisticated spyware. There are plenty of other private companies and, also, governments (including China and Russia) that not only use tools similar to Pegasus but sell them to others.

In 2022 Chinese scientists reported that they had developed "mind-reading" artificial intelligence algorithms that use facial expressions and brain waves to measure a person's loyalty to the Chinese Communist Party. They boasted that the software could be used to "further solidify their determination to be grateful to the party, listen to the party and follow the party." In later chapters I will explain why such algorithms are almost certainly very inaccurate, verging on worthless. The scariest part of this story is that some governments would want to use such algorithms. Even if they don't work, they will surely stifle people's thoughts and behavior and be used to falsely imprison political enemies.

Some might wonder, what harm is there in surveillance if we haven't done anything wrong? Well, for one thing, surveillance turns the government into our enemy. "Of the people, by the people, for the people" becomes "them versus us"—with politicians and bureaucrats doing things to us instead of for us. The election successes of political outsiders (including professional wrestlers, reality show hosts, animals, and the deceased) is partly a reflection of the perception that the government has become a ruling class whose actions are "of them, by them, for them."

Less obvious and perhaps more insidious are the myriad ways in which surveillance, real or imagined, crushes our spirit, our spontaneity, our exuberance. We should feel free to dance like no one is watching and sing like no one is listening. But how can we be truly carefree if we think that we are always being watched and listened to?

An essential part of being human is being ourselves, not being what the government wants us to be. We should sometimes be silly, sometimes sad, sometimes exhilarated. An essential part of being human is being confident that we are entitled to privacy.

The promise of technology is that it will liberate us from mindless tasks. The peril is that the surveillance enabled by technology will enslave us. Instead of computers becoming more like humans, we are in danger of becoming more like computers.

Surveillance Capitalism

It is not just the government spying on us. Governments know a lot about us, but Google and Facebook know even more.

In 1961, President Dwight Eisenhower gave a nationally televised farewell speech in which he warned of the growing power of the military–industrial complex—the alliance between the nation's armed forces and defense contractors campaigning for an arsenal of expensive weapons that makes the military stronger, companies richer, and wars more alluring.

In 2018, Apple CEO Tim Cook invoked memories of Eisenhower when he warned of the growing power of the data–industrial complex—the alliance between government and private businesses to collect and exploit personal data:

Our own information from the everyday to the deeply personal is being weaponized against us with military efficiency. We shouldn't sugarcoat the consequences. This is surveillance.

Big Government and Big Business monitor where we go, what we do, what we buy, and what we write on our computers, e-mail, and social media. Businesses use these data to manipulate us into buying things. Politicians use these data to manipulate us into supporting them. Governments inhibit our behavior by keeping their Big Brother eyes and ears on us.

The costs are not only wasteful purchases, ill-informed votes, and suffocating paranoia but also the diversion of resources from things that really do matter. As a former Facebook data scientist lamented, "The best minds of my generation are thinking about how to make people click ads."

Google

The World Wide Web (WWW) began in the early 1990s as a convenient way for scientists to share information. Soon, everyone wanted to play in the digital sandbox. As the Web grew bigger and more complicated, it became harder and harder to find what one was looking for. In 1994 two Stanford graduate students, Jerry Yang and David Filo, started a website with the fun name "Jerry and David's guide to the World Wide Web," which shared lists they had compiled of interesting web pages, conveniently organized into categories and subcategories within categories.

It was the stereotypical Silicon Valley startup. Filo didn't wear shoes; others wore flip-flops. People coded all-night, powered by sugar and caffeine, and slept in their offices to avoid wasting time traveling to and from their homes. Wagers were settled with shaved heads or tattooed butts. They changed their company's name to Yahoo! and broadcast sassy commercials showing ordinary people using Yahoo searches to find what they wanted, followed by a "Do You Yahoo?" tagline and a maniacal yodel. (You can check out some of these commercials on the Internet, of course, by searching for "Yahoo yodel commercial.") On March 2, 2015, Yahoo celebrated its 20th birthday by breaking the Guinness World Record for Largest Simultaneous Yodel with a webcast of a one-minute yodel by 3432 employees.

In 1996, Yahoo! became a publicly traded corporation when they sold shares to the public at an initial price of $13. By the end of the first day of trading, the price was $33, and the market value of the company was more than Disney, News Corp, and Viacom combined.

Yahoo now had a compilation of 10,000 interesting sites and an average of 100,000 daily Yahoo users. The haystack was exploding and Yahoo hired hundreds of people (nicknamed "surfers") to search the Web for sites. Within months, Yahoo had more than 10 million visitors a day.

Alas, a catchy name and crazy yodeling are not enough to succeed in high tech. This was clearly an unsustainable business model—paying people to spend their days Web-surfing for sites to fit into Yahoo's hierarchical categorization of websites.

Two other Stanford graduate students, Larry Page and Sergey Brin, created a better way. Instead of paying humans to search the Web and guess what pages users might find interesting, they created a search algorithm, called PageRank, that roamed the Web looking for pages and then assessing the importance of each page by counting how many other pages had links to this page. They named their search engine Google, which is an intentional misspelling of googol (the number 10 to the power 100) because the name "fits well with our goal of building very large-scale search engines."

Their insight was brilliant and economical and they offered to sell Google to Yahoo in 1998 for $1 million, so that they could focus on their PhD theses. Yahoo said, *No thanks*. So Page and Brin dropped out of Stanford and raised $1 million to launch Google as a Yahoo competitor.

Four years later, Page and Brin offered to sell Google to Yahoo for $1 billion. Yahoo again said, *No thanks*. Now, Google is a trillion-dollar company and Yahoo is on life support.

Google's search algorithm has morphed into more sophisticated algorithms, allowing Google to dominate the search market. In 2022, it was estimated that Google handled more than 99,000 searches per second, or 8.5 billion searches per day. This enormous database of searchers and search results allows Google to keep improving its algorithms and stay far ahead of would-be competitors. The rich get richer.

Google is not doing this as a public service. This personalized information is used for targeted ads that match companies with the people they want to reach and influence. As they say, when you use a search engine, you are *not* the customer, you are the product.

Interestingly, as idealistic graduate students, Page and Brin wrote a paper describing the Google algorithm and arguing that search engines should be ad-free and non-profit:

This of course erodes the advertising supported business model of the existing search engines. However, there will always be money from advertisers who want a customer to switch products, or have something that is genuinely new. But we believe the issue of advertising causes enough mixed incentives that it is crucial to have a competitive search engine that is transparent and in the academic realm.

So much for youthful idealism. Google has a plethora of products and services and, while we don't know the details of Google's business model, we can be pretty confident that anything that is free is being used to gather data to fuel its ad revenue; for example,

Gmail	every email you send and receive
Google Chrome	every website you visit and what you do there
Google Doc	every word you write
Google Sheets	every number you crunch
Google Slides	every presentation you prepare
Google Sites	everything you put on your website
Google Maps	every place you go

Overall, Google offered nearly 300 products in 2022, with an estimated 1.8 billion people using Gmail and more than 3 billion people using Google Chrome and Google Maps.

Some data collection is not well known. Google advises that "Chrome scans your computer periodically for the sole purpose of detecting potentially un-wanted software," but doesn't publicize the fact that this means Chrome is going through the contents of your files. After secret microphones were found inside Google Nest products, Google's half-hearted defense was that "The on-device microphone was never intended to be a secret and should have been listed in the tech specs."

Google secretly purchased MasterCard transaction data to link ads to sales. Google also tracks sales from sales receipts that are sent to Gmail accounts. (If you go to this page, https://myaccount.google.com/purchases, you can see Google's record of your transactions, along with a not-so-reassuring note that "Purchases made using Search, Maps, and the Assistant are organized to help you get things done, like tracking a package or reordering food."

Google purchased Fitbit, which gives it access to personal fitness data, and Google also obtained millions of medical records (including lab results, medications, and doctor diagnoses) from healthcare providers in 21 states in a secret program named "Nightingale."

Pokémon Go was created at Google and has broad access to your phone data, including your camera and travels. While users are tracking Pokémon, Pokémon Go is tracking them. Plus, local businesses can buy "lures" that pull players to their stores.

Nearly a decade ago, Google executive chair, Eric Schmidt, envisioned this frightening future:

There will be a record of all activity and associations online, and everything added to the Internet will become part of a repository of permanent information . . . . People will be held responsible for their virtual associations, past and present, which raises the risk for nearly everyone since people's online networks tend to be larger and more diffuse than their physical ones.

You Get Used to It

In his movie *Annie Hall*, Woody Allen remembers his childhood living under the big Coney Island roller coaster, Thunderbolt. His inspiration for that fictional recollection was the fact that there actually was a house under the Thunderbolt.

The house came first—built in 1895 as the Kensington Hotel. George Moran bought the hotel in 1925 and, with beachfront land scarce and expensive, decided to have a roller coaster built over the hotel. Figure 3.3 shows that the second floor of the hotel was demolished and steel beams supporting the roller coaster were sent straight through the house, which Moran lived in with his wife Molly.

Figure 3.3 The Thunderbolt house.

After George died, his son moved into the house with a Coney Island waitress and they lived under the Thunderbolt coaster for 40 years. Relatives remember decorative china bolted to the walls, a chandelier that swayed but never broke, and cake and coffee being eaten and drunk while riders screamed a few feet above their heads. They said, "You get used to it after a while."

The property was sold in 1987 and the house was gutted by a fire in 1991. In 2000, New York Mayor Rudy Giuliani demolished the Thunderbolt without consulting the owner, Horace Bullard. A federal jury later ruled that the city had trespassed on Bullard's property, had no justification for tearing down the roller coaster, and had acted with "deliberate indifference." Bullard was awarded $1 million, but refused to accept it because the money would be paid by taxpayers rather than from Giuliani and other city officials.

Why did people live under the Thunderbolt? Why do people choose to live in Siberia? Why do people agree to let tech companies invade their privacy? Why do citizens put up with governments that deny them their fundamental rights to life, liberty, and the pursuit of happiness? Too often, the answer is "You get used to it." Getting used to it may be easier than doing something about it—but it is not a joyful way to live.

The Irony

Science's remarkable successes have improved our lives immeasurably and have deservedly elevated the socioeconomic status of scientists—which can spark envy and even scorn. Their successes have made them targets.

The distrust of science is part of a more general populist backlash against the political, economic, and technocratic elite. Scientists go to good schools, get good jobs, and live well. They often work for governments or are supported by government grants. For many people, distrust of science is part of the distrust of governments that are thought to have too many rules and regulations—including high gasoline taxes, onerous building codes, and mandatory vaccines. To the extent that scientific research is used by governments to justify policies that people dislike, science is viewed as part of the problem.

Science created tools that businesses and governments use to—there is no better word for it—spy on us. They also created the Internet and social media that the most paranoid use to spread conspiracy theories that discredit scientists and sow distrust.

CHAPTER 4

A Post-Fact World

D uring the Watergate hearings in 1974, Indiana Representative Earl Landgrebe boasted that he would not be persuaded by tape recordings or any other evidence implicating President Richard Nixon: "Don't confuse me with the facts. I've got a closed mind." To show that this was not a casual remark or a slip of the tongue, he repeated the statement to reporters and on the *Today* television show. Landgrebe was not the first (or last) person to make this boast—or its closely related cousin, "Don't confuse me with the facts, my mind's made up." Indeed, you can buy t-shirts, coffee mugs, and other merchandise emblazoned with these silly sayings.

Unfortunately, this facts-be-damned attitude is no longer isolated or funny. While she was German Chancellor, Angela Merkel, who has a PhD in physical chemistry from the Berlin Academy of Sciences, warned that evidence-based government policies were threatened by a "post-fact world" in which ideology trumps scientific knowledge.

The World's Only Reliable News

The Sun, a UK tabloid newspaper that publishes all sorts of nonsense, has long included this disclaimer: "SUN stories seek to entertain and are about the fantastic, the bizarre, and paranormal . . . . The reader should suspend belief for the sake of enjoyment." That's pretty clear.

Then there is the American *Weekly World News* (WWN). One distinctive feature is that it is entirely black and white. No color is needed if the stories are colorful enough. Another unusual feature is the combination of a masthead that boasts, "The World's Only Reliable News," with the outrageous stories it prints. When an editor was asked about this, he replied, "People say to me, do you really believe this and that? . . . For heaven's sake, we *entertain* people." As Mark Twain said, "Never let the truth get in the way of a good story"

Interviewed in 1990, this editor explained: "in many cases the *Weekly World News* is the only entertainment they have. . . . Like the guy in rural West Virginia who sees one movie a year and has a black-and-white TV. He gets a real kick out of this paper for 55 cents every week. A lot of people in this country have never heard of countries the serious papers write about, like Nicaragua or Estonia— but *everybody* can relate to a haunted toaster."

One entertaining story was titled, "Alien Orthodontist Returns Home Safely!" The article was about a South Dakota orthodontist who had been abducted by aliens called Gootans and returned to Earth after he straightened their teeth:

The Gootans decided to take Cawley back home. He arrived safely and was reunited with his family and friends. BUT, he is sure he will be abducted again soon. Once the other Gootans see the work I did, I am sure they will all want braces.

Another article offered useful advice: "How to Tell if Your Neighbor is a Zombie":

"People think that if they stay away from the Southern United States or parts of New Mexico, they don't have to worry much about zombies. Unfortunately, that's no longer the case," declares D.C.-based researcher Joh Roskier.

The #1 rule for identifying zombie neighbors is, "You may often notice them gazing blankly into space, even while you're telling them a great story or joke."

WWN seems to be a satire along the lines of *Mad*, *National Lampoon*, or *The Onion*, yet some people take it seriously. A 2010 WWN story about the Los Angeles Police Department planning to spend a billion dollars to buy 10,000 jet packs (watch out Iron Man!) was reported on the *Fox & Friends* morning news show. A 2011 WWN story said that Mark Zuckerberg was shutting down Facebook because it was too stressful for him to manage: "If users want to see their pictures again, I recommend to take them off the internet. They won't be able to get them back once Facebook goes out of business." Some media reported the story as real and did so again when WWN recycled the story in 2012.

There is a market for phony news. Unfortunately, things get messy when people can't tell the difference, or prefer false news to real news.

Benjamin Franklin, Founding Faker

Benjamin Franklin has been described as a man who "merged the virtues of Puritanism without its defects, the illumination of the Enlightenment without its heat." Among his many, many accomplishments were the invention of the lightning rod, bifocals, and the Franklin stove. He also invented a story about British-allied Indians scalping hundreds of American colonists.

After the British army's surrender at Yorktown, the negotiation of a peace treaty began in Paris in April 1782 with Franklin, John Jay, Henry Laurens, and John Adams representing the United States. Hoping for favorable terms, Franklin secretly printed a fake single-page "supplement" (complete with fake advertisements) to the *Boston Independent Chronicle*.

One article in the supplement was a fraudulent letter from an American officer to his commander describing the interception of a shipment of hundreds of human scalps to the British governor of Canada:

struck with Horror to find among the Packages, 8 large ones containing SCALPS of our unhappy County-folks, taken in the three last Years by the Senneka Indians from the Inhabitants of the Frontiers of New York, New Jersey, Pennsylvania, and Virginia, and sent by them as a Present to Col. Haldimand, Governor of Canada, in order to be by him transmitted to England . . . [to] hang them all up in some dark Night on the Trees in St. James's Park, where they could be seen from the King and Queen's Palaces in the Morning.

The letter went on to give grisly details about the hundreds of scalps of soldiers, farmers, women, children, and unborn babies. Package 4, for example, was described as

Containing 102 of Farmers . . . 18 marked with a little yellow flame, to denote their being of Prisoners burnt alive, after being scalped, and their Nails pulled out by the roots and other Torments . . .

Franklin sent the fake supplement to several people and succeeded in having it printed in dozens of newspapers in Europe and the United States.

The American negotiators did, in fact, secure extremely favorable terms from the British, but there is no way of knowing the extent to which those terms were affected by Franklin's shenanigans.

What is certain is that, nowadays, it is much easier reach a wide audience with fake news.

The US of @

The Internet offers the possibility and hope of genuinely free speech with people worldwide able to hear the truth uncensored by government officials—after all, information is power. Those who are oppressed by totalitarian governments would be able to see how others live and be inspired to demand more. Those who are lied to would see the truth.

It hasn't quite worked out that way. Instead, the Internet facilitates the rapid and widespread disseminations of disinformation, both by pranksters and by the malevolent.

Disinfomedia

Jestin Coler grew up in Indiana and settled in a Los Angeles suburb with a wife, two kids, and a minivan. He had always been interested in satirical writing and in 2013 he realized that false news stories were popular on the Internet:

These sites would take a kernel of truth and twist it into a completely fake story to get people all worked up, I got really interested in that, and spent time studying it . . . Ultimately, I decided I wanted to be a part of it as well.

Using the made-up name Allen Montgomery and the semi-plausible site NationalReport.net, Coler and some fellow satire lovers made up fake stories that appealed to an alt-right audience and then watched them take the bait:

The whole idea from the start was to build a site that could infiltrate the echo chambers of the alt-right, publish blatantly false or fictional stories and then be able to publicly denounce those stories and point out the fact that they were fiction.

Soon he realized that not only could he prank the alt-right, he could make money doing it—by selling ads. He created a company called Disinfomedia and, by the time of the 2016 U.S. presidential election, he had 25 writers generating phony news stories for more than two dozen sites and was making about $30,000 a month in ad revenue. The site names sounded legitimate—NationalReport.net, USAToday.com.co, WashingtonPost.com.co, and Denver Guardian ("Denver's oldest news source and one of the longest running daily newspapers published in the United States")—and they reached nearly 100 million page views.

Coler's sites eschewed the alien invasion stories found in the *Weekly World News* but their stories were similarly outlandish:

Muslim Bakery Refuses to Make American Flag Cake for Returning War Veteran
White House Plans To Recruit Illegals To Guard Nation's Border!
President Obama Advocates Eating Dogs In July 4th Address

Coler published a disclaimer, "All news articles contained within National Report are fiction, and presumably fake news." Readers either didn't notice or didn't care. As with WWN, mainstream media circulated some of Coler's fake stories as real. One story said that Arizona was mandating gay-to-straight conversion courses in its schools. Another reported that President Obama was going to personally pay to keep a Muslim Museum open.

Two days before the 2016 election, one of Coler's sites published a story with the headline, "FBI Agent Suspected in Hillary Email Leaks Found Dead in Apparent Murder-Suicide." The story was shared a half-million times on Facebook and viewed by 1.6 million people, prompting NPR to launch an investigation to track down the purported author, Allen Montgomery. Two weeks after the election, they reported that they had found Jestin Coler/Allen Montgomery living in a modest LA suburban home with an unwatered lawn and a large American flag "in a middle-class neighborhood of pastel-colored one-story beach bungalows."

After his identity was revealed, Coler was harassed so badly that he had to change his phone number and move. He shuttered all his sites, except National Report, which now publishes more innocent satire like:

Chris Christie Elected to the Supreme Food Court.
Millions Mourn As Rocker/Activist Ted Nugent, Age 71, Found Alive
35-Year-Old Man Sues His Mother After Being Evicted From Basement Of Her Home

He also changed National Report's slogan "America's *#1* independent news team," to "America's *shitiest* independent news source."

When he wrote an opinion piece for the Nieman Foundation for Journalism at Harvard University, Coler was described as "a recovering fake news publisher."

It turns out there were a lot of Jestin Colers out there, cashing in on the public's apparently unquenchable thirst for phony news. The most successful stories are sensational, but possibly true, and reinforce the beliefs of a large number of people. The FBI agent murder/suicide story confirmed the conviction of many that Hillary Clinton's e-mail leaks were a major crime that had been covered up and that, over the years, several murders and suicides had been connected to the Clintons.

One of the top false news stories that was shared after the election was

CNN: "Drunk Hillary" beat sh*t out of Bill Clinton on Election Night

It had a fake source (CNN) that is widely viewed as liberal and the tantalizing image of Hillary losing it after her unexpected defeat and beating up her womanizing husband. Who wouldn't want to read and share that story?

Paul Horner, Hoax Artist

Paul Horner, dubbed a "hoax artist" by the *Washington Post*, says he made $10,000 a month writing false stories during the 2016 election campaign. His fake stories were intended to be satire, skewering the far right, but were often taken seriously and recycled over the Internet, propelling his hoaxes to the top of Google search results.

Horner's main website was at https://abcnews.com.co, which looked like the real ABC site, except for the .co tacked on at the end. His stories often used the fake byline, Jimmy Rustling, ABC, and always included Horner's name in the tale.

One week before the presidential election, a Horner story appeared with the headline, "Amish in America Commit Vote to Donald Trump, Mathematically Guaranteeing Him a Presidential Victory." The purported author was again Jimmy Rustling, ABC News.

The Amish don't generally vote, but the story nonetheless reported that

"Over the past eight years, the Democratic Party has launched a systematic assault on biblical virtues," said [American Amish Brotherhood] chairman Menno Simons . . . . "Now, they want to put a woman in the nation's highest leadership role in direct violation of 1 Timothy 2:12. We need to stop this assault and take a stand for biblical principles. Donald Trump has shown in both action and deed that he is committed to restoring this country to the Lord's way . . . .

According to statistician Nate Silver of the website fivethirtyeight.com, there are no possible scenarios in which Hillary Clinton can win with Donald Trump carrying the Amish vote . . . .

37-year-old Paul Horner, a self-proclaimed "Donald Trump supporter since day one" told Leilani Hernandez, a reporter with local Columbus news station WBNS-10TV he is thrilled that Trump will be the next President.

". . . I work November 8th and it was going to be tough for me to get off work . . . But now thanks to the great Amish people of this country, they have this thing locked up for a Donald Trump victory against crooked Hillary and I won't have to miss work! God is good!"

. . . The [Amish Brotherhood] chooses weekly charities for those wishing to support what they do. The [Amish Brotherhood]'s charity for the week of October 30th is Sock It Forward, a charity that provides the homeless with brand new socks. If you are interested in learning more about the Amish community and the AAB, you can contact the Pennsylvania Amish Heritage Museum at (785) 273-0325.

The telephone number is one of Horner's favorites, for the Westboro Baptist Church in Topeka, Kansas, which was headed by a flamboyant lawyer named Fred Phelps and has about 70 members, nearly all related to Phelps. Church members regularly picketed funerals, music concerts, and Kansas City Chiefs football games and have been called "arguably the most obnoxious and rabid hate group in America." The church's funeral protests were so obnoxious that several states passed laws specifically banning picketing near funerals; however, in an 8-to-1 decision, the U.S. Supreme Court ruled that "Westboro . . . is entitled to 'special protection' under the First Amendment and that protection cannot be overcome by a jury finding that the picketing was outrageous."

Who Do You Trust?

A study of news stories distributed on Twitter from its inception in 2006 to 2017 used six independent fact-checking organizations (snopes.com, politifact.com, factcheck.org, truthor-fiction.com, hoax-slayer.com, and urbanlegends.about.com) to identify stories as clearly true or false. They concluded that "Falsehood diffused significantly farther, faster, deeper, and more broadly than the truth." While true stories rarely spread to more than 1000 people, the top one percent of false stories were commonly spread to 1000 to 100,000 people.

BuzzFeed compared the 20 top-performing false election stories on Facebook in 2016 with the 20 top-performing mainstream election stories. Figure 4.1 shows that, by November 8, the day of the election, users' engagement (shares, reactions, and comments) with fake news stories outnumbered engagement with mainstream news stories, 8.7 million to 7.3 million.

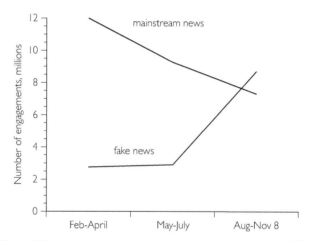

Figure 4.1 Facebook users engagement in election news stories, 2016.

Contributing to the success of false news is a growing distrust of traditional news sources. The Gallup organization has been asking this question for decades:

In general, how much trust and confidence do you have in the mass media—such as newspapers, TV and radio—when it comes to reporting the news fully, accurately and fairly: a great deal, a fair amount, not very much or none at all?

Figure 4.2 shows that in the early 1970s, 18 percent answered "great deal" and only 6 percent said, "none at all." Fifty years later, "none at all" was up to 33 percent and "great deal" was down to 9 percent. For television news, the 2020 survey found that only 18 percent were positive while 76 percent were negative (6 percent had no opinion). In contrast, another survey found that 45 percent of Facebook users trust at least half of the news stories they see on Facebook and 76 percent believe that they can tell the difference between real news and false news.

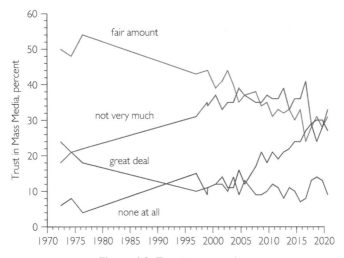

Figure 4.2 Trust in mass media.

Foreign Actors

Three weeks after the 2016 presidential election, *The New York Times* ran a story headlined "Inside a Fake News Sausage Factory," about Beqa Latsabidze, a 22-year-old college student in the former Soviet Republic of Georgia. Early in 2016, as the U.S. presidential campaigns were picking up steam, Latsabidze created a website with the hopes of generating ad revenue from fake stories that praised Hillary Clinton. It flopped. So, he went in the opposite direction, posting fake stories that praised Donald Trump and attacked Clinton. Many of his stories were mildly edited satire from *The Burrard Street Journal* (BSJ), a humorous online news site created by John Egan, a Canadian, which led the *New York Times* to write that

[O]ne window into how the meat in fake sausages gets ground can be found in the buccaneering internet economy, where satire produced in Canada can be taken by a recent college graduate in the former Soviet republic of Georgia and presented as real news to attract clicks from credulous readers in the United States.

Latsabidze's website flourished. One hoax story reported that "the Mexican government announced they will close their borders to Americans in the event that Donald Trump is elected President of the United States." It was the third most-shared story from May to July in 2016.

Latsabidze told the *New York Times* that "For me, this is all about income, nothing more." He had learned through trial and error that Trump supporters "are angry" and "eager to read outrageous tales," so he gave them what they wanted. Latsabidze was hardly the only entrepreneur cashing in on America's voracious appetite for false news.

More than a hundred pro-Trump false-news sites were based in Veles, a town with 45,000 residents in the Balkan nation of Macedonia. The unemployment rate in Veles was 24 percent (close to 50 percent among young people) and, for those who had jobs, average income was less than $400 a month. For many, the lure of making $1000 a month or more by promoting bogus news stories was irresistible.

One young woman justified their actions: "We can't afford anything, and if the Americans can't tell the difference [between real and fake news], it's their fault." Another young Macedonians said, "Teenagers in our city don't care how Americans vote. They are only satisfied that they make money and can buy expensive clothes and drinks!"

Writing in *Wired*, Samanth Subramanian, a London journalist, lamented that

This is the arrhythmic, disturbing heart of the affair: that the internet made it so simple for these young men to finance their material whims and that their actions helped deliver such momentous consequences.

It now seems that the story may be more complicated than this simple narrative. Investigations by BuzzFeed News and others discovered that there were some "secret players" behind the Veles fake-news industry, including some Russians and Americans who were not in it for the money, but to help Trump defeat Clinton.

When Trump ran for re-election in 2020, the Macedonians were back, though trying hard to disguise themselves as ordinary Americans. For example, the About page for the Resist The Mainstream website gives an Austin, Texas, mailing address and says that

Resist the Mainstream is a site for people who have lost complete trust in the mainstream media. Our mission is to report the news the mainstream media won't.

The RTM project was created after its founder experienced heavy censorship by big tech.

Stanford researchers tracked the site back to Veles, where it was run by two Macedonians.

In It for the Money

Americans also have a seemingly insatiable demand for memorabilia, no matter how cheaply it is made or how cheesy it looks. Some are preparing for their nostalgic years; some think they are making a smart investment by acquiring collectibles that will appreciate in value.

The problem is that, as discussed in this book's Introduction, an investment's intrinsic value depends on how much income it generates. Stocks, bonds, rental properties, and profitable businesses all have intrinsic value. Mementos don't. They rely on the Greater Fool theory. If fools don't appear, memento collectors are stuck with cheap and cheesy trinkets.

With the aid of the Internet, deceitful memorabilia hawkers can be based anywhere in the world. A salient example are the many coin-like trinkets featuring Donald Trump's likeness. The Trump coin in Figure 4.3 was clearly modeled after the Kennedy half-dollar in Figure 4.4. The Kennedy coin is legal tender and made of 90 percent silver; the Trump coin is commemorative and said to be gold-plated.

The Trump coin is marketed online on sites denouncing the 2020 election, warning of the dangers of vaccines, and touting the coin as an "Authentic Presidential Re-Election Commemorative Coin." Some sites suggest that the Trump coin may become a cryptocurrency, especially if Trump again becomes president. As I write this, the price on Amazon is $7.95. On one site, the coin is free (plus $9.99 shipping and handling).

One advertisement boasts:

QUALITY CRAFTSMANSHIP: Each coin is gold plated and made to exacting standards to ensure your coin will last for many years to come.

However, among the customer reviews is this 3-star recommendation:

If you decide to get these, I recommend getting multiple and keep the one with the least amount of defects. I bought 4. One has bubbles on the side of his face, another has what looks almost like rust on the edge, and another has a clearly visible scratch on the back and a glob of gunk near the signature.

(I wonder what it would take to get fewer than 3-stars.)

Figure 4.3 The Trump coin.

An investigation by the *New York Times* followed a labyrinth of clues: "What became clear was not just the coin's unusual origins, but an entire disinformation supply chain that relied on falsehoods and misinformation at nearly every step." Among the falsehoods were fake celebrity accounts, including one named RealDenzelWashington that purports to be actor Denzel Washington slamming Democrats:

Democrats are liars! I couldn't handle their agenda and constant lies, so I decided to help team Trump! I turned my attention in the right direction! To the one who deserves it the most, the one who looks after our People and our country. The truth will set us free, and the truth is that President Trump is the best leader for our Country.

Figure 4.4 Kennedy half-dollar.

After the ranting, the account predicts that the Trump coin will soon replace government money.

The *New York Times* gives another example of the depth of the disinformation:

In one post, a fake account for Representative Marjorie Taylor Greene, a Georgia Republican closely aligned with Mr. Trump, shared a fake story on a fake Fox News website about a fake tweet by a fake Elon Musk, falsely claiming that Tesla's chief executive would soon accept Trump coins as payment.

Following lead after lead, the *New York Times* traced the fountainhead of the coin to a Romanian Internet marketing company, Stone Force Media, that, despite all the political ranting and raving, is clearly just in it for the money.

The *New York Times* also found a coin expert with equipment for detecting precious metals in coins:

He found no gold or silver. The coin was also magnetic, suggesting it was mostly made of iron. [Another expert] tested the coin using a nitric acid solution. After he applied a blob to Mr. Trump's gold-colored image, the area darkened, bubbled and then turned green.

"It's paint," he concluded.

What was it worth?

"Nothing," he said.

Commanding the Trend

Cyberwarriors are not in it for the money. In addition to hacking into systems to steal information or bring down networks, cyberwarfare is increasingly focused on using social media to mold people's political views and undermine their faith in society in general and government in particular.

Twitter, Facebook, Instagram, and other social media platforms monitor words, phrases, and hashtags in order to maintain lists of popular ("trending") topics that will attract the attention of users. Government and non-government players, in turn, can use true believers, cyberwarriors, and bots (algorithms that mimic humans) to "command the trend" in order to spread propaganda rapidly to wide audiences.

Air force officer Jarred Prier did a deep dive into cyberwarfare in a 2017 article published in *Strategic Studies Quarterly*. He argued that "He who controls the trend will control the narrative—and, ultimately, the narrative controls the will of the people."

A 2017 study estimated that 48 million Twitter accounts (15 percent of all accounts) are bots. Some bots are used by advertisers; many are malicious attempts to promote false narratives. In 2020 researchers at Carnegie Mellon University studied 200 million tweets about COVID-19 and estimated that between 45 and 60 percent were from bots, and that 82 percent of the most influential COVID-19 retweeters were bots. There is so much misinformation about COVID-19 vaccines that the truth-in-media website NewsGuard has a whole section devoted to refuting COVID-19 myths. As I write this, there are forty-seven myths, including these two whoppers:

> The COVID-19 vaccine causes infertility in 97 percent of its recipients.
> More people have died from COVID-19 vaccines than from the virus itself.

During the 2016 presidential election campaign, Hillary Clinton had her own army of bots trying to command the trend, but Oxford University researchers estimated that pro-Trump bots outnumbered pro-Clinton bots five-to-one.

ISIS Commanded the Trend

The first organization to weaponize Twitter was the Sunni militant group known as the Islamic State of Iraq and Syria (ISIS or IS) that became infamous for posting pictures of executions. The total number of followers was relatively small, so they developed strategies for commanding trends. There were three objectives: project an image of strength to supporters; terrorize opponents; and recruit new fighters and funds.

In 2014 IS unleashed a mobile app called "Dawn of Glad Tidings" that gives users spiritual guidance and updates on IS activities. The real purpose is to create a link to the users' Twitter accounts that automatically retweets IS-branded tweets from the master account. With a sufficient number of retweets, the original tweet trends and catches the attention of users and bots who add to the tidal wave of retweets. At its peak, Dawn of Glad Tidings was generating tens of thousands of retweets a day.

Another effective tactic was trend hijacking. The FIFA World Cup is held every four years and is arguably the most closely followed sporting event in the world. During the 2014 World Cup, IS tweeted thousands of IS propaganda messages using the hashtag #WorldCup2014. One study concluded that, at one point, "nearly every tweet under this hashtag had something to do with IS instead of soccer." Twitter suspended some accounts, but they were quickly replaced with new ones. After @jihadISIS40 was suspended, @jihadISIS41 opened. After @jihadISIS41 was suspended, @jihadISIS42 opened.

Russia Commanded the Trend

Russia has used propaganda to shape the beliefs of its citizens, allies, and enemies for most of the twentieth century, but social media has made it more efficient and effective. Russia has a literal army of professional cyberwarriors who spray a continual "firehose of falsehoods" –a torrent of semi-truths and outright lies. Russia is hardly the only nation engaged in cyberwarfare, but their persistent efforts are instructive.

One of Russia's overarching objectives is to spread distrust that undermines the legitimacy of American media and government. One knowledgeable Russian expert said that they want to make Americans disbelieve everything, because "A disbelieving, fragile, unconscious audience is much easier to manipulate."

Prier found that during racial protests at the University of Missouri in 2015, the trending hashtag #PrayforMizzou included fake stories that the KKK and neo-Nazis were coming to the university campus. Jermaine (@Fanfan1911) tweeted, "The cops are marching with the KKK! They beat up my little brother! Watch out!" and showed an old photo of a badly bruised black child. Approximately 70 Russian bots tweeted and retweeted, pushing the story to the trending list.

The Missouri student body president warned students to stay home and lock their doors. As Prier noted, the hoax went nationwide:

National news networks broke their coverage to get a local feed from camera crews roaming Columbia and the campus looking for signs of violence. As journalists continued to search for signs of Klan members, anchors read tweets describing shootings, stabbings, and cross burnings. In the end, the stories were all false.

While its primary target is the United States, Russian operatives also carry out campaigns in several other countries. After the Missouri disinformation campaign, @Fanfan1911 changed his display name from Jermaine to FanFan and launched a disinformation campaign in Germany with false stories about Syrian refugees, Muslims, the European Union, and German Chancellor Angela Merkel. In the spring of 2016, FanFan started up another misinformation campaign in the United States to disseminate anti-government, anti-Obama, and anti-Clinton stories.

Smear campaigns have been around since the beginning of politics. In the 1800 presidential election contest between Thomas Jefferson and incumbent John Adams, an Adams supporter said of Jefferson:

nothing but a mean-spirited, low-lived fellow, the son of a half-breed Indian squaw, sired by a Virginia mulatto father, as was well known in the neighborhood where he was raised, wholly on hoe-cake (made of course-ground Southern corn), bacon, and hominy, with an occasional change of fricasseed bullfrog, for which abominable reptiles he had acquired a taste during his residence among the French in Paris, to whom there could be no question he would sell his country at the first offer made to him cash down, should he be elected to fill the Presidential chair.

James T. Callender, a Jefferson supporter, published a pamphlet in which he described President John Adams as a "blind, bald, crippled, toothless man who is a hideous hermaphroditical character, which has neither the force and firmness of a man, nor the gentleness and sensibility of a woman." Callender was subsequently jailed for violating the Alien and Sedition Acts and then pardoned when Jefferson was elected president. Denied a position as Postmaster of Richmond, Callender turned on Jefferson, revealing that Jefferson had paid for his attacks on Adams.

Before the Internet, smear campaigns had trouble gaining traction and reaching wide audiences. No more.

After a 2016 speech in which Clinton described some Trump supporters as "deplorables," the word became an Internet rallying cry with users making *deplorable* part of their screen name—including FanFan, who changed his name to "Deplorable Lucy" and used a photo of a white, middle-aged female as his profile picture. It has been estimated that the so-called Deplorable Network of true believers, fake accounts, and bots consisted of approximately 200,000 users.

After the first presidential debate between Clinton and Trump, the top trending hashtag was #TrumpWon, which originated in Saint Petersburg, Russia, the home of the Internet Research Agency, one of dozens of troll farms funded and supported by the Russian government. The Internet Research Agency is known in Internet slang as the "Trolls from Olgino," referring to its offices in Olgino, a historic district of St. Petersburg. They have more than 1000 employees using fake accounts to support Russian political and business interests by spreading fake stories and launching Twitter and Internet trolling attacks throughout the world. The agency supported Russia's annexation of Crimea from Ukraine in 2014 and then focused on helping Trump defeat Clinton in 2016.

A Russian activist named Lyudmila Savchuk got a job at the agency in 2015 and later reported her experiences. There were hundreds of people with jobs like hers, working in two shifts so that fake stories were being pumped out 24/7. Every day, each troll was given a list of ten or so topics to focus on. The United States, European Union, and Putin were always on the list. Sometimes, the posts were subtle nudges; for example, "I am cooking . . . and I had this thought about how bad the [pro-Western] Ukrainian president is." Other posts were more direct; for example, a fake story about a computer game: "created in the States—that even kids loved to play—and the theme of the game was slavery."

Pizzagate

One of the most effective disinformation campaigns was launched to drown out negative reaction to the release of an Access Hollywood tape in which Trump made crude remarks about women. Using Wikileaks, Russian trolls released stolen e-mails from Clinton's campaign chairman John Podesta that were distorted to suggest a variety of scandalous or corrupt activities. The most famous involved a stolen e-mail inviting families to a party at a friend's house with pizza from a store called Comet Ping Pong. This innocent e-mail was deliberately misinterpreted as a coded invitation to a pedophile sex party and circulated through the trending #PodestaEmail hashtag along with a new hashtag, #PizzaGate. Soon, fake stories were circulating, including "A source from the FBI has indicated . . . that a massive child trafficking and pedophile sex ring operates in Washington, D.C. . . . with the Clinton Foundation as a front."

One month after Clinton's election defeat, a 28-year-old father of two from North Carolina decided he would liberate the children who were being abused in the basement of the Comet Ping Pong pizzeria. He drove to the restaurant with an AR-15 semiautomatic rifle, a .38 handgun, and a folding knife, ready to be a hero or die trying. Once inside the restaurant, he shot open a locked door and fired two other shots into a desk and a wall. Behind the locked door were cooking supplies. He couldn't find a basement because there was none.

After his arrest, he conceded that "The Intel on this wasn't 100%," but he did not rule out the possibility of a pedophile sex ring since all he had demonstrated was that there were no children there when he burst in. Some sympathizers suggested that his actions were actually a false flag operation that had been arranged by the Democrat elite to discredit their claims. (The label "false flag" comes from the ruse of disguising one's affiliation; for example, a naval warship that flies the flag of a neutral country.)

Pizzagate soon evolved into QAnon, which extended the conspiracy far beyond the Clintons and found a receptive audience. A 2021 PRRI survey found that 16 percent either completely or mostly agree that, "The government, media, and financial worlds in the U.S. are controlled by a group of Satan-worshipping pedophiles who run a global child sex-trafficking operation."

The Russian Invasion of Ukraine

When Russia invaded Ukraine in 2022, disinformation was a large part of its military strategy. Napoleon observed that "Good intelligence is nine-tenths of any battle." Knowledge of an enemy's strengths, weaknesses, and plans is invaluable. So is deception about one's own capabilities and intentions. As the old adage goes, "In war, truth is the first casualty." A simple example was Russia's insistence that it was carrying out training exercises and its apparent withdrawal from Ukrainian borders shortly before its full-scale assault began.

In modern warfare, disinformation is used to not only deceive the enemy's military but to bolster public support and undermine opposition morale. Thus, Russia promoted the narrative that it had sent a "peacekeeping force" in response to aggression by Ukraine's neo-Nazi leaders—never mind that Ukrainian President Volodymyr Zelenskyy is Jewish and that his great-grandfather and three of his grandfather's brothers died in the Holocaust.

Throughout the war, Russia spread false-flag stories that were intended to rally people against the Ukrainians. For example, Russia claimed that the rapes, torture, and other war crimes committed by Russian soldiers against Ukrainian civilians were carried out by Ukrainians themselves in order to make Russians look bad.

In 2022, the mayors of several European capitals were deceived by deepfake video conference calls purporting to be from Kyiv Mayor Vitah Klitschko. Deepfakes are now used for all sorts of mischief, including fake celebrity pornography and financial scams. Here, computer-generated audio and video made it seem that the mayors were talking to Klitschko. The Berlin mayor said that the person she was talking to "looked exactly like Vitali Klitschko" but, after 15 minutes, she became suspicious when the fake Klitschko began talking about Ukrainian refugees in Germany abusing the welfare system and suggesting that they be sent back to Ukraine to fight in the war.

In one TikTok video, a teary women says that a 16-year-old Russian-speaking boy was beaten to death by Ukrainian refugees in Germany; another reported that a Russian-speaking woman, child, and 70-year-old man were beaten by Ukrainian refugees in Latvia. Both were completely fabricated. A TikTok video that claimed to show a brave Russian soldier parachuting into Ukraine was taken from a 2016 training video.

Ukraine countered with its own stories and videos. When Russia pushed the disinformation that Zelensky had fled Ukraine in anticipation of a crushing defeat, Ukraine posted a video showing Zelensky speaking from the streets of Kyiv. When Russian media claimed that the Ukrainian military had either surrendered or fled in the face of the unstoppable Russian military, Ukraine posted videos of Russian tanks being blown up. When Russia claimed that it only bombed military targets, Ukraine posted videos showing civilian homes, schools, and hospitals that had been destroyed by Russian bombs.

Ukraine also resorted to disinformation. For example, a video purportedly showing six drones putting out a fire in a burning Ukrainian building was actually a Chinese training video. A video showing an 8-year-old Ukrainian girl telling a Russian soldier to "go back to your country" was a 2012 video of a Palestinian girl confronting an Israeli soldier.

It was fitting that a Russian video of a MiG avoiding surface-to-air missiles and a Ukrainian video of a MiG being shot down were both taken from the video game *Arma 3*.

A Post-Truth World

Optimists used to believe that good information would win out over bad information in the court of public opinion. It now appears that the opposite is the case—that we live in a post-truth world where false stories are believed and true stories are dismissed; indeed, the *Oxford English Dictionary* chose "post-truth" as the international word of the year in 2016.

The promise of Facebook, Twitter, Instagram, and other social media platforms is that they can bring people closer together by allowing friends and family to maintain and nourish social ties by sharing events in their lives—a new dog, a birthday party, and even what was eaten for lunch. When Facebook went public, Mark Zuckerberg issued a statement of intent that repeatedly stressed the communal power of sharing:

At Facebook, we build tools to help people connect with the people they want and share what they want, and by doing this we are extending people's capacity to build and maintain relationships.

This was surely corporate BS, and certainly fantastically naive.

One easily foreseen outcome is the allure of posing—exaggerating or diminishing as needed in order to appear happier and more successful than we really are, which provokes others to do the same. I'm having more fun than you. I'm more successful than you. I'm happier than you.

This bragging contest was amped up when Facebook introduced Like and, then, Share buttons and Twitter introduced a Retweet button. The goal changed from sharing and bragging among family and friends to becoming a social media celebrity by "going viral" and amassing followers by posting content that persuades strangers to click buttons. Photos of an expensive party or vacation are not enough; they might even unleash backlash from the internet mob.

Hyperbole, stunts, and dishonesty are rewarded.

We see clickbait that is interesting and supports our beliefs. We click, read, are pleased to have our beliefs confirmed, and share the link with like-minded people. Those with different beliefs are led to different links, which they share with people who agree with them. Instead of nurturing communal feelings, social media is tribalizing in that the world becomes increasingly divided into groups convinced that they are right and others are wrong.

It is very difficult to contain the spread of falsehoods on social media. Twitter could hardly suspend #WorldCup2014 and other popular, legitimate hashtags. Nor was it interested in closing bot accounts since advertisers rely on bot accounts. The trends function could be eliminated, but millions of users appreciate seeing what is hot among the hundreds of millions of tweets sent each day.

Most of the big social media and Internet platforms value engagement above all else. The longer a user is engaged, the more data companies can collect and sell. Unfortunately, one of the most reliable ways to keep people engaged is to feed them sensational falsehoods. In a post-mortem on the 2016 presidential election, former Facebook executive Bobby Goodlatte bemoaned the fact that "As we've learned in this election, bullshit is highly engaging . . . our news environment incentivizes bullshit."

Firehoses of falsehood have been loosely constrained by the number of people doing the dirty work. Now, computer algorithms can generate an essentially infinite supply of disinformation.

In 2021, Apple CEO Tim Cook was blunt:

At a moment of rampant disinformation and conspiracy theories juiced by algorithms, we can no longer turn a blind eye to a theory of technology that says all engagement is good engagement—the longer the better—and all with the goal of collecting as much data as possible.

What are the consequences of seeing thousands of users join extremist groups, and then perpetuating an algorithm that recommends even more? It is long past time to stop pretending that this approach doesn't come with a cost—of polarisation, of lost trust and, yes, of violence.

Outside of social media, false stories are spread through Netflix, Amazon Prime, Hulu, YouTube Video, and other streaming services that offer fake documentaries pushing exaggerations and half-truths. If you watch one, more will be recommended. Keep watching and the recommendations keep coming. Watch enough and it begins to seem like the hogwash is real. A lie told over and over eventually passes for reality. To further sell the lies, fake documentaries are accompanied by fake comments that praise the video and bash the elites who are allegedly trying to suppress the truth.

Trump versus COVID-19

The Trump presidency was a case study of politics clashing with science in ways that undermined the credibility of scientists and endangered ordinary citizens.

When COVID-19 hit the United States in February 2020, the stock market plunged 32 percent, evidently in anticipation of the havoc that would be wreaked on the economy. Until an effective vaccine was developed, the best strategy seemed to be three-pronged: testing, face masks, and social distancing. The Trump administration didn't like any of these. If testing increased, it might reveal how many people were infected. Trump even argued that if there were fewer tests, there would be fewer reported cases—as if that were a good thing. Trump disliked face masks because they conveyed an image of a nation under siege. Going mask-free, as Trump did in public and insisted that his staff do too, projected the desired impression that everything was normal. Trump didn't want social distancing because that would force many businesses to close and weaken the economy—and it is a political truth that voters punish incumbents during economic recessions.

The US Centers for Disease Control and Prevention (CDC) was created in 1946 and has helped win battles against polio, smallpox, measles, and other infectious diseases. It was widely regarded as the world's most trusted public health agency in that it was staffed by expert scientists and free of political influence. All this changed with COVID-19 and the CDC's half-hearted efforts to assert its independence.

A CDC doctor later talked about his frustration as Trump downplayed the seriousness of COVID-19, ridiculed people wearing masks, and promoted hydroxychloroquine and other unproven snake oil. At one point, the doctor was infuriated after seeing a story about an Arizona man who died and whose wife became critically ill after they drank an aquarium cleaner with an ingredient similar to hydroxychloroquine:

I was screaming at the television, "Shut up, you f*cking idiot. Jesus Christ, stop it! You don't know what you're talking about!"

In December 2020, a letter written by a former CDC director was leaked:

The failure of the White House to put CDC in charge, has resulted in the violation of every lesson learned in the last 75 years that made CDC the gold standard for public health in the world. In 6 months, [the White House has] caused CDC to go from gold to tarnished brass.

Throughout the pandemic, Trump officials told CDC officials what they were allowed to say and edited CDC press releases and online messaging to avoid offending Trump.

When the CDC posted online guidelines for reopening schools (including masks, distancing, and staggered schedules), Trump tweeted his objections and threatened to cut off funding to schools that followed the CDC guidelines. Instead of sticking up for scientific integrity, the CDC changed the guidelines. Anyone who has ever known a bully knows that doing what the bully wants encourages the bully to demand more.

One CDC official said that Trump officials constantly checked the CDC website: "If there was a minor tweak to a document to strengthen the recommendations in any way to make it more scientific—they would be notified and told to put it back."

The tension between science and politics peaked shortly before the 2020 presidential election, when Trump officials feared that the stalled economy and rising death toll would cost Trump re-election. On September 13, Michael Caputo, a frustrated Trump-appointed spokesman for the Department of Health and Human Services, said that CDC contained a "resistance unit" of scientists bent on defeating Trump and argued that CDC officials were guilty of "sedition."

The head the CDC at the time was a Trump appointee, Robert R. Redfield, who, when he became director, proclaimed that the CDC was "science-based and data-driven, and that's why CDC has the credibility around the world that it has." However, Redfield's tenure was noteworthy mainly for his unwillingness to push back against political pressures.

For example, during September 15, 2020, testimony before a congressional committee, Redfield said that vaccines would not be widely available until several months after the November presidential election and that face masks might be as effective as a vaccine. Trump was infuriated, telling reporters at a news conference that same day that Redfield had been "confused" and "made a mistake" and that he had called Redfield to let him know that he was wrong. Redfield promptly walked back his comments.

A few days later, William Foege, a former CDC director who is credited with helping eradicate smallpox, sent a blunt letter to Redfield:

[T]his will go down as a colossal failure of the public health system of this country. The biggest challenge in a century and we let the country down. The public health texts of the future will use this as a lesson on how not to handle an infectious disease pandemic.

Foege argued that resigning was not enough:

[Y]ou could send a letter to all CDC employees . . . You could upfront, acknowledge the tragedy of responding poorly, apologize for what has happened and your role in acquiescing, . . . Don't shy away from the fact this has been an unacceptable toll on our country. It is a slaughter and not just a political dispute.

After Trump left office, Dr. Deborah Birx, the White House Coronavirus Response Coordinator, said that a stronger response by the Trump administration could have saved hundreds of thousands of lives. Trump hit back, calling Birx and Anthony S. Fauci, the chief medical advisor to every President since Ronald Reagan, "two self-promoters trying to reinvent history to cover for their bad instincts and faulty recommendations, which I fortunately almost always overturned."

It is no doubt true that Birx and others wanted to rehabilitate reputations that were damaged by their perceived reluctance to contradict Trump during his presidency. On the other hand, it is also clear that Trump was largely focused on the damage to his re-election campaign if the economy were shut down. Trump downplayed the seriousness of the disease and the need for face masks, and promoted unproven treatments, including one bizarre press conference in which he seemed to suggest that people could fight COVID-19 by injecting bleach into their bodies.

A February 2021 report in *The Lancet*, one of the world's top medical journals, concluded that "During his time in office President Trump politicised and repudiated science, leaving the USA unprepared and exposed to the COVID-19 pandemic." A comparison of the U.S. death rate to other advanced countries indicates that

The global COVID-19 pandemic has had a disproportionate effect on the USA, with more than 26 million diagnosed cases and over 450000 deaths as of early February, 2021, about 40% of which could have been averted had the US death rate mirrored the weighted average of the other G7 nations.

The Irony

Page and Brin's 1998 graduate school paper describing the Google algorithm anticipated the inability of search engines to distinguish between fact and fiction and their vulnerability to manipulation by corporations peddling products:

There is virtually no control over what people can put on the web. Couple this flexibility to publish anything with the enormous influence of search engines to route traffic and companies which deliberately manipulate search engines for profit become a serious problem.

They did not anticipate manipulation for political reasons.

The easy access and wide reach of the Internet in general and social media in particular allows pretty much anyone to say pretty much anything and perhaps find a receptive audience, including such evidence-free assertions as the Earth is flat; school shootings are false-flag operations; and Bill Gates orchestrated the COVID19 crisis so that he can use vaccines to insert microchips in our bodies.

Ironically, such far-fetched nonsense is the kind of dragon that science was intended to slay, but now the dragons of fanciful delusion are more powerful than ever because of the Internet and social media that science created and developed. Like a Frankenstein monster that has gotten out of control, the Internet powers the anti-science movement. Too many people have reacted to the heroic successes of scientists in developing safe and effective COVID-19 vaccines with distrust, disinformation, and refusals to be vaccinated.

The costs of rejecting science are enormous, not just for scientists and anti-scientists, but for society as a whole.

PART II

Data Torturing

CHAPTER 5

Squeezing Blood from Rocks

On Friday, October 2, 2020, President Donald Trump revealed that he had tested positive for COVID-19. He was helicoptered to Walter Reed National Military Medical Center that evening and given a battery of experimental treatments. On Monday, he returned to the White House and, a week later, a White House doctor said that Trump had tested negative on consecutive days and was no longer "infectious to others."

Was Donald Trump cured of COVID-19 because he was given remdesivir, dexamethasone, famotidine, a Regeneron antibody cocktail, or a double cheeseburger? Any such conclusion would be an example of the fallacy known as *post hoc ergo propter hoc*: "after this, therefore because of this." The fact that one event occurred shortly after another doesn't mean that the first event caused the second one to happen. It would be a *post hoc* fallacy to credit anything that Trump did or had done to him for the cure.

Such misinterpretations are widespread and seductive. When healthy patients are mistakenly thought to be sick and then "recover" after being given a worthless treatment for a nonexistent illness, it is tempting to conclude that the treatment worked. When someone who is ill fights off the ailment naturally, worthless treatments may again be given credit when no credit is due.

Post hoc fallacies can also happen when people do not recover from an illness. In April 2021, after COVID-19 vaccines were approved for public use, Wisconsin Senator Ron Johnson said that he was "highly suspicious" of the government's vaccination campaign and there was "no reason to be pushing vaccines on people." In May 2021 he argued that the vaccines may not be safe: "We are over 3,000 deaths within 30 days of getting the vaccine."

This is again a *post hoc* fallacy. More than 100 million Americans had been vaccinated at that point and the annual number of deaths in the United States is normally between 8 and 9 people per thousand. Applying that death rate to 100 million people, we expect about 70,000 deaths a month. A more precise estimate would omit deaths from automobile accidents and the like, but would also take into account that the elderly were the first to be vaccinated.

The mere fact that people die does not prove that they were killed by any particular event that happened before their death. Here, we won't know whether a death was due to a vaccination until experts determine whether the vaccine might have played a role. According to the CDC,

CDC and FDA physicians review each case report of death as soon as notified and CDC requests medical records to further assess reports. A review of available clinical information, including death certificates, autopsy, and medical records has not established a causal link to COVID-19 vaccines.

Randomized Controlled Trials

A crucial component of the scientific revolution is the rejection of *post hoc* reasoning and, more generally, an insistence that theories not be accepted until they are tested empirically. In medical research, the most compelling evidence is provided by a *randomized controlled trial* that separates the subjects randomly into a treatment group that receives the medication and a control group that does not. Without a control group, we don't know if changes in the patients' conditions are due to the treatment. Without randomization, we don't know whether the people who choose to take the treatment are younger or healthier, or differ in other ways from those who choose not to take the treatment.

The best medical studies are also double-blind in that neither the patients nor the doctors know which patients are in the treatment group—otherwise, the patients and doctors might be inclined to see more success than actually occurred.

Statistical Significance

There is inherently randomness in randomized controlled trials. Perhaps, by the luck of the draw, there happen to be a larger number of healthy patients whose bodies heal faster in the treatment group, making the medication seem more effective than it really is, or vice versa.

Statisticians handle this problem by calculating the probability that the random assignment of patients would create a disparity between the groups as large (or larger) than what was observed. For example, a study of the effect of vitamin C on the incidence of the common cold was conducted by researchers associated with the University of Toronto during the three-month winter period December 1, 1972, through February 28, 1973. The randomly selected treatment group consisted of 277 subjects who swallowed 1 gram of vitamin C daily and 4 grams on the first day they felt sick. The control group consisted of 285 subjects who followed the same regimen but, unbeknownst to them, consumed inert placebos. Overall, 65 (23 percent) of the treatment group and 52 (18 percent) of the control group reported being free of sickness for the full three months.

The statistical question is the probability that this large a difference might occur by the luck of the draw. Here, there were a total of 562 subjects, of whom 117 were cold-free. If these 562 subjects had been randomly separated into one group of 277 and a second group of 285, what is the probability that the difference in the outcomes would be so large? Statisticians can calculate the answer, which is called the "p-value." Here, the p-value works out to be 0.08.

Is a 0.08 p-value low enough to be persuasive evidence of the effectiveness of vitamin C? Decades ago, a great British statistician named Ronald Fisher set the bar at 5 percent:

It is convenient to draw the line at about the level at which we can say: "Either there is something in the treatment, or a coincidence has occurred" . . . Personally, the writer prefers to set a low standard of significance at the 5 per cent point, and ignore entirely all results which fail to reach this level.

Fisher has been lauded as "a genius who almost single-handedly created the foundations for modern statistical science." When he chose a p-value cutoff of 0.05, that became gospel—no matter that he later acknowledged that other cutoffs might sometimes be more appropriate.

By Fisher's 5 percent rule, the 0.08 p-value in this Canadian study is not statistically significant.

The Hydroxychloroquine Hoax

In March 2020, as the reality of COVID-19 was taking hold, an op-ed column in the *Wall Street Journal* co-authored by the chair of the National Advisory Commission on Rural Health and the director of the Division of Infectious

Disease at the University of Kansas Medical Center reported that French researchers had treated COVID-19 patients with a combination of the malaria drug hydroxychloroquine and a Z-Pak (antibiotic), and that every single patient was cured within six days. In comparison, the recovery rates were 57.1 percent for patients treated with hydroxychloroquine alone and only 12.5 percent for patients who received neither medication. More tests were needed, but the op-ed argued that we should treat first, test later. What harm could that do?

Well, for one, there may be unknown side effects. In addition, if people begin stockpiling hydroxychloroquine, less will be available for those who suffer from malaria, rheumatoid arthritis, and other serious conditions that the medicine is intended to treat. A less obvious problem is that a false sense of security might seduce some people into behaving less cautiously and increase the number of infections dramatically.

Here, there were several warning signs. First, only six patients had been treated with hydroxychloroquine and Z-Pak. The best response should not be "Wow, you found a cure!" but "Wow, not many patients in that study!" Second, the people who received hydroxychloroquine were, on average, 14 years younger than those who did not. The different outcomes may well have been due to the age differences.

Third, the author of the French study revealed that

Six hydroxychloroquine-treated patients were lost in the follow-up during the survey because of early cessation of treatment. Reasons are as follows: three patients were transferred to intensive care unit . . . one patient stopped the treatment . . . because of nausea . . . one patient died.

Wait, what? Shouldn't the research question be whether the drug helps patients leave the hospital alive?

Fourth, when a field is hot, there is a race to publish. It is hard to imagine a field hotter than COVID-19 in 2020 and, indeed, 200,000 COVID-19 papers were published that year. Citizens and governments wanted a cure and scientists wanted to be heroes. An avalanche of rubbish was inevitable. With a large number of researchers trying a wide variety of treatments and rushing to publish, it's almost certain that someone somewhere will find a coincidental pattern that is statistically significant but meaningless. The only solution, alas, is further testing.

Real Science Trumps Wishful Thinking

Nonetheless, President Trump was soon touting hydroxychloroquine as a miracle cure, boasting that it had "a real chance to be one of the biggest game changers in the history of medicine." In May 2020, he told reporters that he himself was taking it every day. In a June press conference, Florida governor Ron DeSantis announced that he had ordered a million doses for Floridians and, to demonstrate its effectiveness, he showed a video clip of a patient who had taken hydroxychloroquine and recovered from COVID-19 (an obvious *post hoc* fallacy).

In July, Trump retweeted a video of Dr. Stella Immanuel's claim that she has cured hundreds of COVID-19 patients with hydroxychloroquine: "You don't need a mask. There is a cure." Her credibility might be slightly dampened by some of her other claims—that alien DNA is used in medical treatments, some government officials are reptiles, and some health problems are caused by sexual relations with witches and demons during dreams. Nonetheless, Trump said she was "spectacular," "very respected," and an "important voice." Donald Trump, Jr., proclaimed the video a "must watch."

Scientific evidence is more than anecdotes, wishful thinking, and mustwatch videos. Real science uses randomized controlled trials. Well-run trials have now been done and the results are in. Here is a report from one clinical researcher:

I have spent almost the entirety of the last 5 months caring for patients . . . Most have survived. Many have not. All have received a huge number of drugs and other interventions. It would be impossible for me to know, for a particular patient, if they lived or died because of hydroxychloroquine, or one of the other therapies I administered, or because they were younger/older or had more medical problems, or just because they had a better/worse case . . . . The only way to know if a medication works is a clinical trial, and thankfully we did them . . . a lot of them . . . and now we know. Hydroxychloroquine does not work for COVID-19.

Several other trials reached the same conclusion—hydroxychloroquine is ineffective and has potential side effects, including possible heart problems.

In the immortal words of Nobel Laureate Richard Feynman, "If it disagrees with experiment it is wrong. In that simple statement is the key to science . . . . If it disagrees with experiment it is wrong. That is all there is to it."

The claim that hydroxychloroquine prevents or cures COVID-19 is wrong. That is all there is to it.

Real Science

The rapid development of safe and effective COVID-19 vaccines was an absolutely stunning achievement, but it wasn't done with sloppy science. It was done with real science.

The first confirmed COVID-19 infections were reported in Wuhan, China, on December 31, 2019. The first U.S. case was someone who had traveled to Wuhan and returned to the state of Washington on January 15, 2020. The World Health Organization (WHO) issued a Global Health Emergency warning on January 31, and, in February, said it did not expect a vaccine to become available for at least 18 months.

It took only 11 months.

While politicians squabbled, scientists did what scientists do. They got to work to develop a safe and effective vaccine.

One very large obstacle was the financial risk for pharmaceutical companies. Research and large-scale testing would require hundreds of millions, perhaps billions, of dollars and there was no guarantee that the companies would get a return on their investment. It might take years to develop and test a vaccine. What if mask-wearing and other factors eradicated the virus before vaccines were ready? What if a company did not find an effective vaccine, or another company beat them to market?

Determined to get a vaccine approved before the November 3 presidential election, the U.S. government committed $14 billion to support the pharmaceutical companies' vaccine efforts. Pfizer, a 170-year-old pharmaceutical company, teamed up with BioNTech, a 12-year-old German biotechnology company, and declined to accept federal money. It risked $2 billion of its own money—a risk mitigated by the federal government agreeing to spend $1.95 billion to buy 100 million doses if Pfizer developed a vaccine approved by the FDA—regardless of whether the vaccine was still needed. This was subsequently increased to $5.87 billion for 300 million doses.

Moderna, on the other hand, a 10-year-old American biotech company with no approved products, was given $954 million for research and development and guaranteed a federal purchase of 300 million doses for $4.94 billion.

Johnson & Johnson, another old-time big Pharama company, was given $456 million for research and development and promised $1 billion for 100 million doses. Four other companies were also supported but, as of July 2022, had not received approval for their vaccines. Worldwide, more than 300 COVID vaccine projects were launched although, as of June 2022, only three vaccines had been approved in the United States and another six in Europe.

The search for a safe, effective vaccine was not based on *post hoc* reasoning or wishful thinking. Pfizer and Moderna decided to go with a promising but unproven type of vaccine called messenger RNA (or mRNA) that, in essence, teaches cells how to make a harmless spike protein that triggers the production of antibodies needed to fight COVID-19. The theory behind mRNA had been around for decades in work involving hundreds of researchers, but no mRNA drug or vaccine had ever been approved by the FDA.

The seemingly insurmountable problem was that the body's natural immune system detected the alien RNA and destroyed it before it could complete its mission. After years of frustration, Katalin Karikó and Drew Weissman, two University of Pennsylvanian researchers who are now Nobel Prize candidates, figured out a way for mRNA to slip through our bodies' natural defenses, and their work inspired the founders of BioNTech and Moderna. Indeed, the name Moderna is a combination of *modified* and *RNA*.

Other companies decided to use previously proven approaches that rely on an inactive virus to produce an immune response. In every case, the vaccines were based on sound scientific knowledge. What was not known was how safe and effective the vaccines would be. The FDA was aiming for 50 percent effectiveness.

After its development, the Pfizer mRNA vaccine was tested in a randomized controlled trial of 37,586 participants, with 18,801 people in the treatment group given the vaccine and 18,785 people in the control group given an injection of a saline placebo. There were 8 cases (1 severe) of COVID-19 in the treatment group, compared to 162 (3 severe) in the control group. Instead of the hoped-for 50 percent effectiveness, there was a 95 percent reduction in infections, from 162 to 8. The possible side effects included temporary pain, headache, and tiredness. Similar tests of Moderna's mRNA vaccine yielded similar results.

The FDA approved Pfizer's Emergency Use Authorization request on December 11, 2020. Moderna's vaccine was approved a week later and then Johnson & Johnson's vaccine. Soon, millions of doses were being distributed.

Figure 5.1 shows the weekly number of worldwide deaths attributed to COVID-19, which totaled 6.65 million as of December 12, 2022. These official statistics do not include people who did not test positive for coronavirus before dying—which can be quite large in countries with little COVID testing.

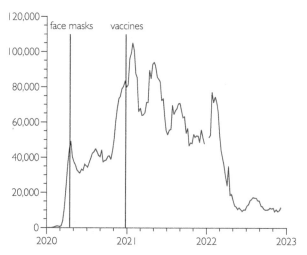

Figure 5.1 Number of deaths worldwide attributed to COVI-19.

The first drop in COVID-related deaths occurred after the CDC's April 3, 2020, recommendation that people wear face masks. The second drop came after vaccines became available. There were subsequent surges in deaths related to the spread of variants of COVID and then a decline as more and more people became vaccinated. As of December 2022, the unvaccinated were 17 times more likely to die of COVID than were those who had been given a vaccination and booster.

Real science really works.

P-Hacking

Fisher's argument that we need to assess whether empirical results can be explained by simple chance is compelling. We don't want to misinterpret coincidences as real effects.

However, any hurdle for statistical significance (including Fisher's 5 percent rule) is bound to become a target that researchers strive mightily to hit. Remember Fisher's language:

the writer prefers to set a low standard of significance at the 5 per cent point, and ignore entirely all results which fail to reach this level.

No researcher wants their findings to be ignored entirely, so many will work hard to get their p-values below 0.05. Publish or perish is real—promotion, funding, and fame hinge on being published. If journals require statistical significance, researchers will give them statistical significance.

This is called *p-hacking*—trying different combinations of variables, looking at subsets of the data, discarding contradictory data, doing whatever it takes until something with a low *p*-value is discovered and then pretending that this is what you were looking for in the first place.

A p-hacker might begin by considering the data as a whole. Then look at males and females separately. Then look at different races. Then differentiate between children and adults; then between children, teenagers, and adults; then between children, teenagers, adults, and seniors. Then try different age cutoffs. Let the senior category be 65+; if that doesn't work, try 55+, or 60+, or 70+, or 75+. Eventually, something clicks and a low *p*-value is discovered. As Ronald Coase, an Economics Nobel Laureate, cynically observed: "If you torture data long enough, they will confess." Stephen M. Stigler, a renowned statistics professor added the punchline, "but confessions obtained under duress may not be admissible in the court of scientific opinion."

The undermining of the *p*-value target is an example of Goodhart's law (named after the British economist Charles Goodhart) which states that "When a measure becomes a target, it ceases to be a good measure." Goodhart was an economic advisor to the Bank of England and his argument was that setting monetary targets causes people to change their behavior in ways that undermine the usefulness of the targets. We now know that Goodhart's law applies in many other situations as well. For example, when the Soviet Union's central planning bureau told nail factories to produce a certain number of nails, the firms reduced their costs by producing the smallest nails possible—which were also the least useful. Setting the target undermined the usefulness of the target.

Here, the scientific demand for evidence that is statistically significant has, ironically, undermined the usefulness of the evidence that scientists report.

In 1983 Ed Leamer, a professor of management, economics, and statistics at UCLA, wrote a visionary paper with the cheeky title "Let's Take the Con Out of Econometrics." His argument was that economists and other researchers almost invariably engage in trial-and-error "specification searches" involving a large number of plausible models before choosing the results they report. He identified the problem and warned of the dangers of p-hacking before p-hacking was even a word.

A study of psychology journals found that 97 percent of all published test results were statistically significant. Similar results were found in a study of medical journals. Surely, 97 percent of all the tests that researchers conduct do not yield statistically significant results, but too many journal editors believe that results that are not statistically significant are not worth publishing. A rejection letter from a major medical journal revealed this attitude quite clearly:

Unfortunately, we are not able to publish this manuscript. The manuscript is very well written and the study well documented. Unfortunately, the negative results translate into a minimal contribution to the field.

The same attitude reigns outside academia. A business or government researcher trying to demonstrate the value of a strategy, plan, or policy feels compelled to present statistically significant empirical evidence.

With these compelling incentives, researchers chase statistical significance. With large amounts of data and fast computer algorithms, it is an easy pursuit. On average, even if researchers just analyze correlations among randomly generated data, one out of every twenty correlations, on average, will be statistically significant at the 5 percent level.

Pressured to produce statistically significant results, researchers individually and collectively test an unimaginable number of theories, write up the results that have low p-values, and discard the rest. The problem for society is that we see the statistically significant results but not the tests that didn't work out. If we knew that behind the reported tests were zillions of unreported tests and remember that, on average, one out of every twenty tests of worthless theories will be statistically significant, we would surely view what does get reported with well-deserved skepticism.

Some researchers seem blissfully unaware of the perils of p-hacking. Daryl Bem, a prominent social psychologist, encouraged researchers to torture their data:

Examine [the data] from every angle. Analyze the sexes separately. Make up new composite indexes. If a datum suggests a new hypothesis, try to find further evidence for it elsewhere in the data. If you see dim traces of interesting patterns, try to reorganize the data to bring them into bolder relief. If there are participants you don't like, or trials, observers, or interviewers who gave you anomalous results, place them aside temporarily and see if any coherent patterns emerge. Go on a fishing expedition for something—anything—interesting.

Using the fishing-expedition approach, Bem was able to report evidence for some truly incredible claims, such as "retroactive recall": People are more likely to remember words on a recall test if they study the words *after* they take the test.

In 2011 Bem was given a Pigasus Award by master skeptic James Randi. The name *Pigasus* is a combination of the word *pig* and the mythological winged horse *Pegasus*, which together refer to the cynical expression, "when pigs fly." Randi described the award this way:

The awards are announced via telepathy, the winners are allowed to predict their winning, and the Flying Pig trophies are sent via psychokinesis. We send; if they don't receive, that's probably due to their lack of paranormal talent.

When others redid Bem's experiments, they found nothing at all. No word yet on whether Bem predicted or received his Flying Pig award.

The Pizza Papers

In 2016 Brian Wansink was a Professor of Marketing at Cornell and Director of the Cornell Food and Brand Lab. He had written (or co-authored) more than 200 peer-reviewed papers and two popular books, *Mindless Eating* and *Slim by Design*, which have been translated into more than 25 languages.

Wansink became famous peddling the alluring message that anyone can lose weight without draconian diets or exhausting exercise but, instead, by simple, painless tricks: "The best diet is the one you don't know you're on."

On an ABC news show, Wansink said that people eat less if their food is served on small plates. On the Rachel Ray television show, he said people eat less if their kitchens are painted with neutral earth tones instead of bright or dim colors. He also advised that the color of dinnerware and tablecloths affect how much people eat. Such optimism was an easy sell, not only to lazy dieters, but esteemed academics. In their book *Nudge*, economics Nobel laureate Richard Thaler and Harvard Law School professor Cass Sunstein described Wansink's experiments as "masterpieces."

Then Wansink wrote a blog post titled, "The Grad Student Who Never Said No," about a student who had been working with data that had been collected at an all-you-can-eat Italian buffet and not getting statistically significant results. Wansink reassured the student that "I don't think I've ever done an interesting study where the data 'came out' the first time I looked at it." He advised her to separate the diners into

males, females, lunch goers, dinner goers, people sitting alone, people eating with groups of 2, people eating in groups of 2+, people who order alcohol, people who order soft drinks, people who sit close to buffet, people who sit far away, and so on . . .

Then she could look at different ways in which these subgroups might differ: "# pieces of pizza, # trips, fill level of plate, did they get dessert, did they order a drink, and so on . . ." He concluded that she should "Work hard, squeeze some blood out of this rock."

By never saying no to p-hacking, Wansink and the student co-authored four papers that are now known as the "pizza papers." The most famous one reported that men eat 93 percent more pizza when they dine with women.

By boasting publicly about his p-hacking, Wansink invited scrutiny. Nick Brown took the bait. P-hacking is typically difficult to prove, so Brown focused instead on more direct and obvious errors in the pizza papers, including conflicting, incorrect, implausible, and impossible statistical results.

Wansink refused repeated requests to make his data public so that Brown and others could check his calculations. He argued unconvincingly that someone might use the data to figure out the identities of the pizza eaters—imagine how embarrassing that might be! Andrew Gelman, a professor of statistics and political science at Columbia University, responded:

It seems pretty simple to me: Wansink has no obligation whatsoever to share his data, and we have no obligation to believe anything in his papers.

Brown and others were able to reconstruct the data that could have generated the reported statistical results. They found that, not only were the implied data often implausible, but the reported statistical results were frequently contradictory or impossible.

In January 2017 Brown and two collaborators—Jordan Anaya, a computational biologist in Virginia, and Tim van der Zee, a graduate student at Leiden University in the Netherlands—posted a paper online titled "Statistical Heartburn: An Attempt to Digest Four Pizza Publications from the Cornell Food and Brand Lab" that detailed 150 apparent errors in the four pizza papers:

The sample sizes for the number of diners in each condition are incongruous both within and between the four articles. In some cases, the degrees of freedom of between-participant test statistics are larger than the sample size, which is impossible. Many of the computed F and t statistics are inconsistent with the reported means and standard deviations . . . . We contacted the authors of the four articles, but they have thus far not agreed to share their data.

They also began blogging about their findings. Brown later recalled that "At one point I wrote three blog posts in as many weeks and someone commented, 'Ah, I see it's Wansink Wednesday.'" In February *New York Magazine* published an article titled "A Popular Diet-Science Lab Has Been Publishing Really Shoddy Research." Other national media picked up the story and Wansink and Cornell were forced to respond.

Wansink sent e-mails with the subject line "Moving forward after Pizza Gate" to dozens of colleagues, complaining of "cyber-bullying" and protesting that the only problems were some "missing data, rounding errors, and [some numbers] being off by 1 or 2."

After a quick internal review, the Cornell administration concluded that "while numerous instances of inappropriate data handling and statistical analysis in four published papers were alleged, such errors did not constitute scientific misconduct." A skeptic might think that Cornell's main objective was to protect the reputation of the university and one of its media stars. After all, what do two psychology graduate students and a computational biologist know about food research?

Meanwhile, suspecting that the pizza papers might be the proverbial tip of an iceberg of sloppy science, Brown and others began scrutinizing dozens of other Wansink papers and discovering mistake after mistake. Van der Zee started a Wansink Dossier web page for cataloging the errors people found. Soon, serious errors had been found in forty-five publications.

Brown's attention had been drawn to the pizza papers by the p-hacking boasts, but he was astonished by the large number of basic errors in the papers he looked at: "The level of incompetence that would be required for that is just staggering." Either much of the data had been made up, or someone had, over and over again, been very sloppy when describing the research and recording the statistical results.

One paper reported that elementary school children "ranging from 8 to 11 years old" were swayed to eat fruit instead of cookies by decorating the fruit with stickers of Elmo from Sesame Street. Brown was deeply dubious. How many 11-year-olds would be seduced by Elmo stickers? Brown was right, with Wansink eventually admitting that "We mistakenly reported children ranging from 8 to 11 years old; however, the children were actually 3 to 5 years old."

In another study, this time claiming that elementary school students were more likely to eat carrots that were labeled "X-ray Vision Carrots," the number of students in the study was sometimes said to be 113, sometimes 115, and sometimes 147. The reported average number of carrots that the children took (17.1) was not equal to the sum of the average number of carrots they ate (11.3) and didn't eat (6.7).

P-hacking e-mails also surfaced. In one, a co-author reported that "I have attached some initial results ... Do not despair. It looks like stickers on fruit may work (with a bit more wizardry)." Wansink fretted that one p-value was 0.06: "If you can get the data, and it needs some tweeking, it would be good to get that one value below .05." For another paper, Wansink wrote his co-authors that "there's been a lot of data torturing with this cool data set," adding in another e-mail that "I think it would be good to mine it for significance and a good story."

Gelman concluded that "to repeatedly publish papers where the numbers don't add up, where the data are not as described: sure, that seems to me like research misconduct." Others emphasized the p-hacking more than the numerical errors. Susan Wei, a biostatistician at the University of Minnesota, said that "He's so brazen about it, I can't tell if he's just bad at statistical thinking, or he knows that what he's doing is scientifically unsound but he goes ahead anyway."

Cornell was forced to reopen its investigation and, after ten months, concluded that Wansink had committed academic misconduct. Wansink seemed surprised, proclaiming that "from what my coauthors and I believed, the independent analyses of our data sets confirmed all of our published findings." Nonetheless, he handed in his resignation.

Wansink has acknowledged that there were "some typos, transposition errors, and some statistical mistakes," but he has persisted in arguing that "There was no fraud, no intentional misreporting, no plagiarism, or no misappropriation." He seemingly continues to view p-hacking as a virtue, not a vice.

In December 2019, Wansink posted an essay titled "Research Opportunities to Change Eating Behavior," which summarized eighteen of the papers that had been retracted, with suggestions for how others might redo these studies. To facilitate such re-research, he included assessments on a scale of 1 to 5 of the potential practical usefulness and potential effort required for each of these eighteen studies. There were also several color photographs of the restaurants and participants in his original studies. The explanations for the retractions included such obvious errors as one paper reporting the mean ages of the participants when no age data were collected and other papers reporting empirical results that were not consistent with the raw data. To date, I don't believe that anyone has followed in Wansink's muddy footsteps.

Unwitting P-Hacking

In medical testing, a false positive occurs when a person who does not have a particular disease is diagnosed as having it; for example, when a person who does not have COVID-19 tests positive. In statistical testing, a false positive occurs when there is no real relationship but the results are deemed statistically significant. Fisher's 5 percent rule is intended to set the probability of a false positive at 0.05 but p-hacking can push the false-positive probability much higher.

Researchers sometimes p-hack without even recognizing it. For example, a researcher might do an experiment with twenty subjects and get a p-value of 0.06, disappointingly close to, but above, the magical 0.05 threshold. Knowing that more data often give lower p-values, the researcher tests another ten subjects and, if this doesn't do the trick, tests another ten. If and when the p-value drops below 0.05, the researcher stops the experiment, satisfied that statistical significance has been achieved.

The problem with this seemingly reasonable procedure is that if the stopping point is not specified ahead of time, the probability of a false-positive result is larger than 0.05, perhaps much larger. In fact, if the sample size were allowed to grow indefinitely, it is essentially certain that the p-value will drop below 0.05 at some point. No one has the resources to do an experiment with an unlimited sample size, but the probability of a false positive increases surprisingly fast with modest flexibility in the sample size.

To demonstrate this, I generated two independent sets of random variables, so that any statistically significant correlation between them is necessarily a false positive. Figure 5.2 shows what happens to the false-positive probability when an initial sample of twenty subjects is augmented by an additional ten subjects, up to a maximum of 200 subjects.

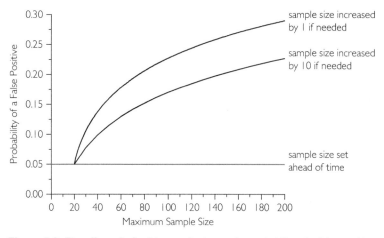

Figure 5.2 The effect of a flexible sample size on the probability of a false-positive outcome.

If the sample size is fixed ahead of time, the probability of a false positive is always 5 percent, no matter what the sample size. If, however, the sample size starts at twenty and is increased by ten as needed, the probability of a false positive increases to 17 percent if the researcher is willing to go to 100 observations and 23 percent if the maximum is 200.

Figure 5.2 also shows that a strategy of increasing the sample size one observation at a time (instead of ten) increases the chances of a false-positive outcome even more. A willingness to go up to a maximum of 200 observations, one additional observation at a time, increases the false-positive probability to 29 percent.

I was surprised by the dramatic effect of adding observations and I expect that most researchers who have used this strategy would be surprised, too. One reported that

We ran subjects in chunks and checked the effect along the way. It was something like 25 subjects run, then 10, then 7, then 5. Back then this did not seem like p-hacking. It seemed like saving money.

A flexible sample size is not the only trick for getting p-values below 0.05. The data might contain *outliers* that are very different from the rest of the data. Sometimes, an outlier shows up in the data because of a clerical error that can be corrected. For example, an age might mistakenly be recorded as 450 instead of 45. Other times, the data are recorded correctly and yet some values are substantially different from the rest. For example, a study of middle-class lifestyles might include a billionaire who lives, well, very differently. If the research is truly focused on the middle class, the billionaire can be rightfully removed from the sample.

There is no mathematical rule for identifying data as outliers, so it is ultimately a subjective decision to be made by researchers—which provides welcome flexibility for those seeking to get their p-values below 0.05.

The relationship shown in Figure 5.3 has a tantalizing p-value of 0.051. It is understandable that a researcher who wants the data to confirm a relationship between these two variables will be tempted to delete the observation labeled as a possible outlier. If this observation is omitted the p-value drops to 0.043, which is statistically significant. Mission accomplished!

Both of these strategies—adding additional data and omitting inconvenient data—seem reasonable, even justified. However, they are both p-hacking and they both increase the chances of a false-positive conclusion. To demonstrate this, Figure 5.4 shows the effect of supplementing the experiment reported in Figure 5.2 with an additional p-hack—at every stage, delete an observation in order to reduce the p-value. The omission of more than one outlier would make

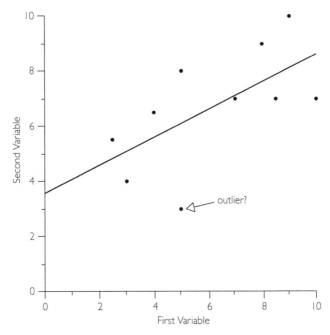

Figure 5.3 Omitting data can lower p-values.

it even easier to get the p-value below 0.05, and increase the probability of a false-positive conclusion even more.

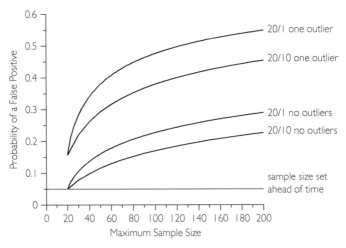

Figure 5.4 Two p-hacks are better than one.

Election Models

Presidential election models are entertaining but often unreliable in that they predict past elections astonishingly well and then do poorly with new elections—leading to adjustments that allow the models to continue to predict the past, but not the future.

For example, on the eve of the 2020 U.S. presidential election, when Joe Biden was a heavy favorite to defeat Donald Trump, it was widely reported that a model created by Helmut Norpoth, a political science professor at Stony Brook University, predicted that Donald Trump had a 91–95 percent chance of winning re-election and most likely would win 362 electoral votes to Joe Biden's 176 votes. The model was based on three flexible statistics: a presidential voting cycle, long-term trends in partisanship, and performance in the primaries. For elections prior to 1952, the primaries variable incorporated all presidential primaries; for the 1952–2004 elections, only the New Hampshire primary was used; and for the 2008–2020 elections, the New Hampshire and South Carolina primaries were used. Norpoth seemed utterly unaware of the dangers of p-hacking his models to fit past data.

His university newspaper reported that

While some might suspect that unusual circumstances—e.g., the COVID-19 pandemic and the civil unrest in the wake of the George Floyd killing—might have an unpredictable effect on the election results, Norpoth said those crises have no bearing on his projection.

"My prediction is what I call 'unconditional final,'" he said. "It does not change. It's a mathematical model based on things that have happened.

After Biden defeated Trump by 306 to 232 electoral votes, Norpoth was unbowed, attributing his wild miss to "a perfect storm" of events that he had previously dismissed as irrelevant:

the outbreak of the coronavirus pandemic; then, triggered by the lockdown aimed to keep people safe, came an economic downturn of a scale not seen since the Great Depression; then, with the same goal in mind, came an unprecedented expansion of voting by mail; add to that the killing of George Floyd, which sparked a wave of racial unrest not seen since the 1960's.

Another widely reported model was constructed by Alan Lichtman, a history professor at American University. On the eve of the 2016 presidential election, he was virtually alone in predicting that Donald Trump would defeat Hillary Clinton. His model used thirteen true/false questions ("keys") and was said to have predicted the winner of the popular vote in every presidential election since 1984. The thirteen keys are sensible political and socioeconomic issues; for example,

No primary contest: There is no serious contest for the incumbent party nomination.
Strong short-term economy: The economy is not in recession during the election campaign.
No social unrest: There is no sustained social unrest during the term.

When 8 or more of the thirteen keys are true, the incumbent party is predicted to win the popular vote.

After Trump's surprising victory (even Trump seemed surprised, as he had not prepared a victory speech and was planning to fly to Scotland to play golf), Lichtman's "correct" forecast was widely reported, even though it was not correct—Lichtman had predicted that Trump would win *the popular vote* and he lost to Hillary Clinton by nearly 3 million votes.

One problem with Lichtman's model is what statisticians call *overfitting*. There were eight presidential elections between 1984 and 2012 and it is a mathematical fact that the winner in each of these eight elections can be explained perfectly by any seven independent explanatory variables such as whether an American League team wins the World Series, whether it is raining in Porterville, California, and whether a Japanese scientist is awarded a Nobel Prize in Chemistry. Notice that I used the word *explained* instead of *predicted* because finding variables that fit the past is not really predicting. Lichtman's model with thirteen variables is overkill.

A second problem is that Lichtman has p-hacked his model over time to better explain the past. Lichtman's original model, reported in 1981, had twelve keys. The current model dropped one key ("Has the incumbent party been in office more than a single term?"), added two foreign policy/military keys, and changed one key (from "Did the incumbent party gain more than 50% of the vote cast in the previous election?" to "After the midterm elections, the incumbent party holds more seats in the U.S. House of Representatives than after the previous midterm elections"). The large number of changes after a handful of election results—three of which (1964, 1972, and 1984) were easily predicted landslides—is a sure sign that the model is better at predicting the past than forecasting the future.

A third problem is that the model has considerable wiggle-room in that some of the true/false questions are extremely subjective, which allows Lichtman to adjust his predictions as needed. For example, Lichtman waited an unusually long time to make his 2020 prediction for the contest between Donald Trump and Joe Biden. The pollsters overwhelming predicted a Biden victory and Lichtman finally did too, stating that only six keys were true, two short of the required eight keys. But consider these two keys which Lichtman counted as false:

Foreign/Military Success, because of the lack of an acclaimed success abroad.
Incumbent Charisma, because Trump appeals only to a narrow base

Lichtman counted Foreign/Military Success as false even though Trump was nominated for a Nobel Peace Prize for his work in facilitating the Abraham Accords, a peace agreement between Israel and the United Arab Emirates. Was that achievement worthy of a *true* checkmark? Who knows? That's the problem.

The Incumbent Charisma key is even more problematic. Who could seriously deny that Trump has charisma? The rationalization that "Trump appeals only to a narrow base" sounds suspiciously circular: Trump appeals to a minority of voters, so I predict that he will lose.

Models constructed to fit the past are not reliable predictors of the future. You can seldom see where you are going by looking in a rear-view mirror.

The Cult of Statistical Significance—and Neglect of Practical Importance

Fisher's 5 percent rule has become the default hurdle for statistical significance. A completely separate issue is whether the results are of any practical importance. In the Canadian vitamin C study, the 0.08 p-value was insignificant (by Fisher's rule). A separate question is whether an increase from 18 to 23 percent in the chances of going cold-free is substantial. Is that difference worth the cost of vitamin C and the possible side effects? Some may answer *yes* and some may answer *no*, but not because of the p-value.

One casualty of the adoption of Fisher's criterion (or, indeed, any criterion) for statistical significance is the shoving aside of the question of practical importance. Too many researchers report results that are statistically significant but of trivial importance.

What Did You Do in the 1960s?

In the 1960s, while others were frolicking in legal and illegal ways, a man named Willard H. Longcor was busy enjoying a lifelong hobby—recording the results of dice rolls. His preferred method was to roll a single die and jot down a 1 or a 0 to indicate whether the outcome was an even or odd number.

He contacted a famous Harvard statistician (Frederick Mosteller) and offered to share the fruits of his hobby. Longcor subsequently rolled 219 dice 20,000 times each, a total of 4,380,000 rolls, of which 2,199,714 were even numbers and 2,180,286 odd numbers. The p-value for a test of the presumption that even and odd numbers are equally likely is about 1 in a hundred million trillion, but the observed difference is arguably of little practical importance. Even numbers came up 50.2 percent of the time, which, to me, seems trivially different from 50 percent. You might disagree, but that is exactly the point. Practical importance is a subjective opinion which has nothing whatsoever to do with statistical significance.

Oomph

The importance of practical importance has long been recognized. In the 1930s, William S. Gosset, who derived the t-test that is widely used to gauge statistical significance, noted that agricultural researchers would like to be able to say that "We have significant evidence that if farmers in general do this they will make money by it." More recently, in 2016 the American Statistical Association declared that "A p-value, or statistical significance, does not measure the size of an effect or the importance of a result."

Deirdre McCloskey has sharply and repeatedly criticized economists for prioritizing statistical significance over practical importance (what she calls "oomph"). The charge is serious, and the language is sometimes vivid; for example, she wrote that "Most of what appears in the best journals of economics is unscientific rubbish."

She and Stephen Ziliak looked at empirical articles published in the 1980s and 1990s in the *American Economic Review*, which is considered to be one of the most prestigious economics journals, and concluded that most of the papers were flawed and that the situation has gotten worse over time. Specifically, they reported that 70 percent of the papers published in the 1980s did not distinguish between statistical significance and practical importance and that 82 percent of the 1990s papers made the same error. A widespread focus on statistical significance rather than practical importance has also been found in journals in the fields of psychology, education, and epidemiology.

Ananya Sen, Claire Van Note, and I analyzed 306 empirical papers that were published during the years 2010–2019 in *MIS Quarterly*, generally considered to be one of the top information systems journals. We found that 78 percent of the papers used statistical significance alone to judge the success of their models.

For example, one study investigated the factors affecting German household decisions to adopt smart metering technology (SMT) for monitoring electricity consumption. One factor they considered was household income: "Consumers with higher income are able to spend on environmental friendly devices such as SMT and are more likely to adopt it," but the researchers do not say how much more likely—they do not consider the practical importance of their findings. Nor did their paper give enough information for readers to judge for themselves.

The authors wrote that "Intention is the subjective probability that a person will perform a certain behavior;" however, I contacted the authors and learned that their intention variable is not a probability but, instead, each household's average response to three questions on a scale of 1 to 7. The income variable is defined as "the average income of the consumers," but we are not told whether it is weekly, monthly, or annual income and whether the data are recorded in euros, thousands of euros, or some other unit. It turns out that the data are monthly income measured in thousands of euros.

This detective work yields our oomph answer: a 1000-euro increase in monthly household income is predicted to increase a household's intention to adopt SMT on a 1-to-7 scale by a trifling 0.062, from an average of 4.125 to 4.187.

The Irony

The concept of statistical significance was developed by scientists to assess whether empirical results can be explained away by coincidence. However, a myopic focus on statistical significance fuels tenacious p-hacking and models judged by p-values alone are apt to fare poorly with fresh data.

One very big cost is that smart, hard-working people waste their time p-hacking. A second cost is that when we are told that a medication, business practice, or government policy has statistically significant effects and then find that the effects vanish or are trivial when the recommendations are followed, our disappointment erodes the credibility of scientific research.

CHAPTER 6

Most Medicines Disappoint

C hapter 5 explained why randomized controlled trials are the gold standard for medical tests. However, a positive result in a randomized controlled trial does not guarantee that a medical treatment is effective. After all, randomization, by definition, means that there is some randomness in the results and, so, there is always a chance that a positive result is a false positive.

Such disappointments are magnified by the reality that the number of worthless treatments that can be tested is much larger than the number of genuinely useful treatments. Table 6.1 shows a stylized example. Ten thousand treatments are tested, of which 500 (5 percent) are genuinely useful. A randomized controlled trial is assumed to identify correctly 95 percent of all useful treatments as useful and 95 percent of all useless treatments as useless. Thus, 475 of the tests of useful treatments (correctly) show statistically significant effects and 9025 of the tests of useless treatments (correctly) do not show statistically significant effects.

With all these 95s floating around, it might seem that 95 percent of the tests with statistically significant results are useful. Paradoxically, this is not so. The 475 false positives are as numerous as the 475 true positives. Half of all certified useful treatments are, in fact, worthless.

The odds are even worse if more than 95 percent of all tested treatments are worthless. Adding to the noise is the temptation researchers have to p-hack the data to find statistically significant relationships. Even though the reported p-value is less than 5 percent, the true false-positive rate may be much higher than that.

Table 6.1 *True Positives and False Positives*

	Significant at 5 percent level	Not Significant	Total
Useful treatment	475 (true positive)	25	500
Useless treatment	475 (false positive)	9025	9500
Total	950	9050	10000

Suppose, for example, that p-hacking increases the probability of a false-positive result on any single test from 5 to 10 percent. If so, the appropriate adjustments of the numbers in Table 6.1 show that two-thirds of all positive test results are false positives. If p-hacking increases the false-positive rate on a single test to 20 percent, the overall false-positive probability increases to 80 percent.

John Ioannidis, who holds positions at the University of Ioannina in Greece, the Tufts University School of Medicine in Massachusetts, and the Stanford University School of Medicine in California, is a medical-research whistle-blower. His 2005 paper, "Why Most Published Research Findings Are False," attracted worldwide attention to the false-positive problem in medical research. He followed up his theoretical argument with a study of forty-five of the most widely respected medical studies. He found that attempts to corroborate the original studies were only made in thirty-four cases and were only success-ful in twenty. Of these thirty-four highly respected studies that were retested, fourteen (41 percent) were apparently false positives.

In addition to p-hacking, another reason why positive test results may be exaggerated is that medical trials typically focus on patients who are known to have a specific illness. Outside the trials and inside doctors' offices, treatments are often prescribed for patients who have different illnesses or a combination of illnesses. For example, antibiotics are widely viewed as a miracle drug and are often very effective. However, some doctors seem to prescribe antibiotics reflexively despite possible side effects that include allergic reactions, vomiting, or diarrhea. The *ICU Book*, the widely respected guidebook for intensive care units, advises that "The first rule of antibiotics is try not to use them, and the second rule is try not to use too many of them."

Given all these difficulties, it is not surprising that many "medically proven" treatments fall short of expectations. The pattern is so commonplace in medical research that it has a name—the "decline effect."

One reflection of the fragility of many medical studies is that the medical profession keeps changing its collective mind. Chocolate was once bad for you; now dark chocolate is good for you. Coffee was once blamed for all sorts of cancers; now, if anything, coffee reduces the risk of cancer. Butter was once bad for you; now butter is healthy and margarine is unhealthy. Eggs were once bad for you; now eggs are okay. This medical mind-changing undermines the credibility of medical researchers. I personally suffered through decades of chocolate chip cookies made with margarine instead of butter. I am not happy about that.

Some people are so confused and jaded by medical flip/flops that they think scientists are withholding effective treatments for cancer and other diseases so that charlatans can peddle worthless nostrums. A profession's reputation can't sink much lower than that.

The decline effect is not limited to a handful of researchers. In 2019, it was reported that 396 of the 3017 randomized clinical trials published in three premier medical journals during the past dozen years were medical reversals that concluded that previously recommended medical treatments were worthless, or worse. Many of the reversals were for widely used procedures. For decades, hormone replacement therapy was recommended for menopausal women to prevent cardiovascular disease and other chronic ailments. Now, research indicates that there are minimal benefits and increased risk of strokes, blood clots, and cancer. Arthroscopic surgery is currently used hundreds of thousands of times each year to treat patients with cartilage tears and osteoarthritis of the knee. It turns out that surgery is no more effective than regular physical therapy.

In 2011 an analysis by the Bayer pharmaceutical company of published data for sixty-seven in-house projects with positive results found that two-thirds were not confirmed by subsequent analyses. In 2012, the biotechnology firm Amgen reviewed fifty-three landmark cancer studies and found that only six had been confirmed in subsequent studies. A venture capitalist has reported that the "unspoken rule" among early stage venture capitalists is that at least half of published drug studies, even those in top-tier academic journals, "can't be repeated with the same conclusions by an industrial lab."

Shortcuts

In 2016, the U.S. Congress passed the 21st Century Cures Act which is intended to accelerate medical development by allowing the Food & Drug Administration (FDA) to bypass clinical trials and, instead, approve medical drugs, devices, and treatments based on "real world evidence" like insurance claims, anecdotes, and other observational data.

The bill's title, 21st Century Cures Act, suggests this is a modern approach to medicine, designed to replace outmoded procedures. The label "real world evidence" suggests that randomized controlled trials are artificial and obsolete.

Thus, the FDA recently implemented a "precertification program" that eliminated randomized controlled trials for smartphone health apps and other health software. A 2021 FDA report with a tongue-twisting title, Artificial Intelligence/Machine Learning (AI/ML)-Based Software as a Medical Device (SaMD) Action Plan, explains the premise:

Artificial intelligence (AI) and machine learning (ML) technologies have the potential to transform health care by deriving new and important insights from the vast amount of data generated during the delivery of health care every day . . . . One of the greatest benefits of AI/ML in software resides in its ability to learn from real-world use and experience, and its capability to improve its performance. FDA's vision is that, with appropriately tailored total product lifecycle-based regulatory oversight, AI/ML-based Software as a Medical Device (SaMD) will deliver safe and effective software functionality that improves the quality of care that patients receive.

In later chapters, I will explain why AI has overpromised and underdelivered in many areas, including medicine. Here, I want to focus on the superficially appealing idea that, instead of randomized controlled trials for assessing effectiveness and safety, medical software can be approved based on modest criteria and then tested continuously with real world data.

It could be argued that software can help us and can't hurt us—so why fool around with lengthy and expensive clinical trials? Apple watch's irregular heart rhythm notification provides a cautionary answer. Irregular heart rhythm can be a sign of atrial fibrillation which can lead to a variety of heart-related problems including strokes and heart failure.

Apple watches (commonly referred to as iWatches) have an electrical heart rate sensor that can measure the wearer's heart rate and heart rhythm. When Apple's heart app was launched, the President of the American Heart Association described it as "game-changing."

In 2017 Apple launched the Apple Heart Study, an $8 million two-year collaboration with Stanford University that monitored 419,297 volunteers who turned on the irregular rhythm notification app on their iWatches.

Of the 219,179 volunteers under the age of forty, a total of 341 people were notified of irregular rhythms, but only nine were confirmed as having atrial fibrillation. At the other end of the age spectrum, there were 24,626 people who were sixty-five or older, of whom 775 received a notification, but only sixty-three were confirmed as having atrial fibrillation.

Unfortunately, many important questions went unanswered. We don't know the false-positive rate, because many of the people who received notifications simply ignored them. We don't know the false-negative rate because the people who did not receive notifications were not tested. In addition, there was surely self-selection bias in the people who own iWatches and volunteered to participate and in the people who received notifications and ignored them.

So, how useful is the Apple Watch heart app? The unwelcome truth is that no one knows. The only way to answer the questions that need answering is with the randomized controlled trials that the FDA's precertification program is designed to circumvent. Part of the FDA's original mission was to protect consumers from untested and possibly harmful treatments that the public is in no position to evaluate. The FDA's precertification program is a step in the wrong direction.

One thing we do know is that software manufacturers have enthusiastically endorsed the precertification program because it allows them to sell lots of untested apps—digital snake oil.

The 21st Century Cures Act reflects and stokes the distrust of science in two quite different ways. First, the law itself suggests that the country's elected leaders don't believe in randomized controlled trials. Second, the inevitable disillusionment with untested nostrums will further undermine the public's faith in science.

Doctors Won't Be Obsolete Anytime Soon

The decline effect is likely to be pervasive when computer algorithms are involved because computers are so good at storing data and calculating correlations and so bad at judging whether the data are relevant and the correlations make sense.

As the very old proverb goes, "There's many a slip twixt the cup and the lip." The possible slip-ups begin with the patient data, which are notoriously unreliable. A careful analysis of British hospital records found an annual average of

1600 reports of adults over the age of thirty using *child and adolescent* psychiatry services and a comparable number of reports of people under the age of twenty using *geriatric* services. Tongue firmly in cheek, the authors speculated that "We are not clear why so many adults seem to be availing themselves of pediatric services, but it might be part of an innovative exchange program with pediatric patients attending geriatric services."

They also found thousands of reports of men using outpatient obstetrics, gynecology, and midwifery services each year, though there were fewer reports of women availing themselves of vasectomies. These were clearly clerical errors.

Electronic medical records (EMRs) have imposed some standardization on patient data collection, though this well-meaning effort to record all relevant information has created what has been called *iPatient medicine*, with voluminous data logged in for every patient visit and procedure. The excessive collection of medical data has been dubbed "hypermetricosis," a disease not of the patient but of the health care system.

A 2019 report from the National Academy of Sciences estimated that, on average, doctors and nurses spend 50 percent of their workday interacting with their computer screen instead of their patients. Doctors complain of burnout from EMR overload, but it is now what is expected, indeed required, of them. A survey of emergency room doctors found that it typically took six clicks to order an aspirin, eight clicks for a chest X-ray, and fifteen clicks for a prescription. On average, more than four hours of a ten-hour emergency room shift was spent entering data in computers, even though the label *emergency* suggests there is some urgency in treating patients.

EMRs are intended more for standardizing the billing process than for improving patient care. Check boxes can trigger payments but often cannot adequately convey the important details of a doctor's observations. One doctor complained that "There might be 8, 9 pages or more on something you could categorize in 2 or 3 pages. It's numbing, mind numbing. It's ream after ream of popup screens and running data." Another said that the written patient notes that preceded EMRs could document a patient's condition and treatment in ways that the doctor and other doctors could easily understand.

But now I get a 14-page note, and if you don't know what I did, you'd have to really dive through 14 pages to find what should be an easy evaluation. We're creating mountains of data that make more noise than pertinent fact. I think the computer is a great tool, but I went into medicine to work with people and not to be a data entry clerk.

EMRs can be processed easily by computer software, but that doesn't make them more useful. Whether entering patient information on a clipboard or in a computer, an age of 80 will occasionally be recorded as "8" and vice versa; and the *male* box will sometimes be checked when *female* was intended. It doesn't happen often, but it is important that the diagnoses and treatments not be led astray when it does happen. Human doctors will recognize such mistakes and correct them. Computer algorithms won't because they have no idea what age and sex mean, which may lead to incorrect and possibly dangerous diagnoses and treatments.

Using computer algorithms to give diagnoses and recommend treatments is dodgy enough without adding EMR errors to the mix. An expert told me that

Physicians are well aware of the limitations of EMR and routinely work around them. . . . I am not saying that EMRs are unsafe. However, the problems in EMR data are major challenges when using these data with computer algorithms.

There are similar problems with a computer's digital library of healthcare research. Multiple diseases are the norm in the real world, particularly with elderly patients, and people who are ill often take multiple medications. This reality undercuts the usefulness of a computer-retrieved medical study that is restricted to patients who have one disease and take one medication.

In his book *The Digital Doctor*, Robert Wachter wrote that

One of the great challenges in healthcare technology is that medicine is at once an enormous business and an exquisitely human endeavor; it requires the ruthless efficiency of the modern manufacturing plant and the gentle hand-holding of the parish priest; it is about science, but also about art; it is eminently quantifiable and yet stubbornly not.

In 2020, a team of physicians and biostatisticians at the Mayo Clinic and Harvard reported the results of a survey of clinical staff who were using computerized clinical decision support (CDS) to improve glycemic control in patients with diabetes. When asked to rate the success of the computer system on a scale of 0 (not helpful at all) to 100 (extremely helpful), the median score was 11. The most common complaints were that the recommended treatments were not sufficiently tailored to each patient, the recommendations were inappropriate or not useful, and the system inaccurately classified standard-risk patients as high-risk.

Walmart is now planning a major expansion into healthcare, based on Clover Health, which boasts that its Clover Assistant technology "gives your primary care doctor a complete view of your overall health and sends them care recommendations that are personalized for you, right when you're in the appointment," using "data, clinical guidelines, and machine learning to surface the most urgent patient needs." That sounds more like advertising copy written by a marketing executive than reliable advice written by doctors. Walmart shoppers may want to be wary of the bargain healthcare aisle.

There have been many great successes in medical science: the HIV triple-drug combination that has transformed HIV from a death sentence to a chronic condition, the benefit of statins, the effectiveness of antibiotics, the treatment of diabetes with insulin, and the rapid development of COVID-19 vaccines. There have also been many successes in identifying the causes of diseases: asbestos can lead to mesothelioma, benzene can cause cancer, and smoking is associated with cancer.

Even though medical research has enjoyed many wonderful successes, blind faith is not warranted. Researchers should bear in mind the reasons for the decline effect and doctors and patients should anticipate the decline effect when deciding on treatments.

Exercise is Medicine

The United States spent $3.8 trillion on healthcare in 2019, before COVID-19. That's 18 percent of U.S. Gross Domestic Product and nearly $13,000 per person. That's more than double the average spending in dozens of comparable countries, yet U.S. healthcare outcomes are near the bottom of any list. Americans have the highest obesity rates in all age groups and the second highest death rate from heart disease. For life expectancy at birth, the United States ranks thirty-fourth, behind Chile and Lebanon.

The United States has arguably the best doctors, medicines, and hospitals. Why are so many Americans in poor health and why do so many die young? We can blame the system or we can blame ourselves.

Robert Sallis was born in Whittier in the 1960s and grew up in Brea, two sleepy Southern California cities near Buena Park and La Habra, the two drowsy cities where I grew up. Bob's father was a school teacher/coach and his mom was the office manager of a family-medicine medical group—a combination that sparked Bob's lifetime interest in sports and medicine.

Basketball was his passion and, though he wasn't tall (a tad under 6 feet), he was intelligent on and off the court. During his senior year in high school, he set the Southern California high school record for assists in a single game and a season. Meanwhile, Gregg Popovich, who had been the Air Force Academy's top scorer and was good enough to be invited to try out for the 1972 U.S.

Olympic team, was now an assistant coach there, and he was looking for good players who were smart enough to meet the Academy's tough entrance requirements. Bob was successfully recruited and played basketball for four years at the Air Force Academy. Bob clearly wasn't going to have a career as a professional basketball player, and Coach Pop encouraged him to pursue his other passion by going to medical school, which he did.

After Sallis graduated from the Air Force academy and medical school, he returned to Southern California and a residency at Kaiser Permanente in Fontana, California. Coach Pop was now the head basketball coach at Pomona College in nearby Claremont and he invited Bob to be the team doctor, which he happily accepted. Medicine and sports were his passions and he was now immersed in both. Coach Pop soon left to become the future Hall-of-Fame coach of the San Antonio Spurs, but Bob continued working with Kirk Jones, Pomona's Athletic Trainer and Professor of Physical Education, who happened to keep meticulous injury reports on Pomona College athletes. For example, one highly cited paper that they and I co-authored concluded that there is "very little difference in the pattern of injury between men and women competing in comparable sports."

In another study, Bob and a research assistant found that patients with mild to moderate Parkinson's Disease who participated in a cycling endurance training program significantly improved their condition.

Working with medical patients and athletes, Bob became increasingly convinced of the health benefits of regular exercise. He is a past-president of the American College of Sports Medicine and currently chairs Exercise is Medicine, a joint initiative of the American College of Sports Medicine and the American Medical Association. He has also served as chair of the Healthcare Sector for the National Physical Activity Plan and is a spokesperson for Every Body Walk!, a national initiative designed to get America up and walking.

He believes that "physical activity is like a wonder drug, useful for both the treatment and prevention of virtually every chronic disease." He argues that if exercise was included in the Physician's Desk Reference (the largest compendium of drugs), it would be the most powerful drug currently available—and it would be malpractice for doctors to not prescribe it.

Several empirical studies support his arguments. It is well known that immune function improves with regular physical activity and that those who are regularly active have a lower incidence, intensity of symptoms, and mortality from various viral infections. Regular physical activity improves cardiovascular health, increases lung capacity, increases muscle strength, improves mental health, and has beneficial effects on multiple chronic diseases.

Bob and a Kaiser research team now have additional evidence, based on a statistical analysis of 48,440 Kaiser Permanente patients who had been infected with COVID-19 and also had at least three measurements of their physical activity level during the two-year period prior to their infection. The U.S. Physical Activity Guidelines recommend an average of 150 minutes a week of moderate to vigorous physical activity (like a brisk walk). That's just a 30-minute walk five days a week, yet only 3118 of these patients (6.4 percent) met this guideline during the two years preceding their infection. An astonishing 6984 patients (14.4 percent) reported 0 to 10 minutes a week of moderate to vigorous physical activity! Most of these consistently inactive patients had no exercise at all. The remaining 38,338 patients (79.1 percent) were in-between the extremes of consistently inactive and meeting the guidelines.

Bob and his research team found that, controlling for dozens of possible confounding factors including medical conditions that the CDC has identified as associated with an increased risk of severe illness from COVID-19, the patients who met the physical activity guidelines were far less likely to be hospitalized, admitted to intensive care units, or die. Physical inactivity was a larger risk factor for death than were cardiovascular disease, diabetes, and other chronic diseases—and physical inactivity is something we have control over. Some physical activity was better than none at all, but best of all is meeting the guidelines of 150 minutes a week.

In a paper published in the *British Journal of Sports Medicine*, the world's leading peer-reviewed journal in sports medicine, they concluded that

Physical inactivity is the strongest modifiable risk factor … We recommend that public health authorities inform all populations that short of vaccination and following CDC guidelines, engaging in regular physical activity may be the single most important action they can take to prevent severe COVID-19 and its complications, including death.

Yes, exercise really is medicine and not just for COVID-19.

So, launch yourself out of your chair and go for a walk!

The Irony

Medical advances have been among the greatest scientific successes, including germ theory, anesthesia, vaccines, antibiotics, X-rays, and organ transplants. Scientists also instituted the randomized controlled trials that allow them to obtain reliable estimates of the benefits and side effects of proposed medical treatments.

However, randomized controlled trials are generally gauged by statistical significance, which encourages the p-hacking that is partly responsible for the disappointing performance of many "proven-effective" treatments and the resulting disillusionment with medical research.

CHAPTER 7

Provocative, but Wrong

S
ome scientists do research for the pure love of science. The simplicity of the Pythagorean theorem ($a^2 + b^2 = c^2$) that relates the lengths of the sides of a right triangle is remarkable. The appearance of π in so many equations describing natural phenomena is amazing. The universal gravitational constant, speed of light, and Planck's constant are astonishing. There is something deeply satisfying about mathematical and scientific discoveries.

Others take pleasure in research that benefits others. The development of vaccines for polio, COVID-19, and other diseases is immensely gratifying. So were the inventions of the internal combustion engine, telephone, and light bulb. James Tobin, an economics Nobel Laureate, spoke for many when he wrote that

I studied economics and made it my career for two reasons. The subject was and is intellectually fascinating and challenging, particularly to someone with taste and talent for theoretical reasoning and quantitative analysis. At the same time it offered the hope, as it still does, that improved understanding could better the lot of mankind.

Tobin's research on identifying the causes of economic recessions and the ways that governments might combat them is part of a macroeconomic corpus that has literally saved the world from economic catastrophe.

There can also be fame and fortune. Who wouldn't want to be as celebrated as Darwin, Einstein, or Edison? Realistically, a minuscule number can. Instead, we settle for 15 minutes of fame.

The easiest way to get media buzz is to report provocative findings—that the fate of the stock market depends on who wins the Super Bowl, that hurricanes are deadlier if they have female names, that traffic accidents surge on Friday the

13th, that people eat less if their plates, cups, and kitchens are certain colors—which is an inducement to do what it takes to get titillating results that will be reported widely. Recognizing this incentive, whenever I encounter provocative research, my default assumption is that it is wrong.

The *BMJ* Christmas Issue

The *British Medical Journal* (BMJ) is one of the world's oldest and most prestigious journals. Each Christmas, they take time off from the usual dry academic papers and publish studies that are noteworthy for their originality: "We don't want to publish anything that resembles anything we've published before." Although the papers are unusual, *BMJ*'s editors assure readers that

While the subject matter may be more light-hearted, research papers in the Christmas issue adhere to the same high standards of novelty, methodological rigour, reporting transparency, and readability as apply in the regular issue. Christmas papers are subject to the same competitive selection and peer review process as regular papers.

The articles are often goofy, and four have won the dreaded Ig Nobel Prizes for research that is trivial and laughable:

> *Side effects of sword swallowing.* There can be problems if the swallower is distracted or the sword has an unusual shape.
>
> *People with acute appendicitis driving over speed bumps.* Speed bumps are more painful for people who have acute appendicitis.
>
> *MRI imaging of male and female genitals during sexual intercourse.* Move along; nothing to see here.
>
> *The effect of ale, garlic, and soured cream on the appetite of leeches.* This experiment was abandoned for ethical reasons after two leeches died from exposure to garlic.

Some articles are clearly meant not to be taken seriously. For example, one article proposed that researchers who receive rejection letters from academic journals respond by rejecting the rejection. The authors include a handy template, a snippet of which is shown here:

Dear Professor [insert name of editor]
[Re: MS 2015_XXXX Insert title of ground-breaking study here]
Thank you for your rejection of the above manuscript.

Unfortunately we are not able to accept it at this time. As you are probably aware we receive many rejections each year and are simply not able to accept them all . . . . Please understand that our decision regarding your rejection is final. We have uploaded the final manuscript in its original form, along with the signed copyright transfer form. We look forward to receiving the proofs and to working with you in the future.

The *BMJ* Christmas issue is widely anticipated, read, and reported in the mainstream media, which leads some researchers to p-hack in myriad ways and demonstrates how flawed research can slip past even the best journal editors. Here are several examples.

Scared to Death

A 2001 *BMJ* Christmas paper argued that many Japanese and Chinese consider four to be an unlucky number because the pronunciation of "four" and "death" are very similar in Japanese, Mandarin, and Cantonese. Based on this assertion, the authors made the outrageous argument that Japanese and Chinese Americans are so terrified by the number four that they are prone to heart attacks on the fourth day of every month. The article was titled, "The Hound of the Baskervilles Effect," in reference to the Sir Arthur Conan Doyle's story in which Charles Baskerville dies while being pursued by a vicious dog down a dark alley:

The dog, incited by its master, sprang over the wicket-gate and pursued the unfortunate baronet, who fled screaming down the yew alley. In that gloomy tunnel it must indeed have been a dreadful sight to see that huge black creature, with its flaming jaws and blazing eyes, bounding after its victim. He fell dead at the end of the alley from heart disease and terror.

The claim that the fourth day of the month is as terrifying to Asian-Americans as being chased down a dark alley by a ferocious dog is so preposterous (and racist) that I knew it had to be flawed. Sure enough, it was.

The number of coronary deaths for Japanese and Chinese Americans is essentially the same on the third, fourth, and fifth days of the month. The Baskervilles authors ignored this inconvenient truth and, instead, separated the deaths into several categories and reported the numbers for categories that had more deaths on the fourth than on the third or fifth. In addition, the data were restricted to the years 1989 to 1998 even though data for 1969 through 1988 were available.

When I retested their claim using their cherry-picked categories and the 1969–88 data they omitted, I found that there was nothing special about the 4th. (A fuller discussion is in my book, *Standard Deviations: Flawed Assumptions, Tortured Data, and Other Ways to Lie With Statistics*.)

A subsequent study led by Nirmal Panesar, a medical professor at Chinese University of Hong Kong, noted that the number four is unambiguously unlucky for Cantonese, but not so for Japanese and Mandarin-speaking Chinese and, also, that the numbers fourteen and twenty-four are even more ominous than the number four because fourteen sounds like "must die" and twenty-four sounds like "easy to die." Panesar also pointed out that people who believe that these are deadly numbers are more likely to follow the lunar calendar than the Gregorian calendar used in the Baskervilles paper.

Panesar analyzed coronary deaths among the Chinese population of Hong Kong, of whom 95 percent speak Cantonese and are arguably more superstitious than Americanized Asians about the numbers four, fourteen, and twenty-four. Even using the cherry-picked heart disease categories, there were not significantly more deaths on the fourth, fourteenth, or twenty-fourth days of the month using either the Gregorian or lunar calendar. They concluded that "perhaps the hound was barking up the wrong tree."

What the p-hacking hound had done was sniff out parts of the data that supported a ludicrous conclusion.

Unhappy Birthdays

The 2020 *BMJ* Christmas issue included a study reporting that surgeries are more likely to be fatal if they are done on the surgeon's birthday. It is a truly damning indictment of doctors if patients are dying because surgeons are distracted by birthday plans and good wishes from their assistants. However, several red flags were waving and I investigated the claim.

The study involved patients sixty-five or older who underwent one of seventeen common surgeries between 2011 and 2014: four cardiovascular surgeries and the thirteen most common non-cardiovascular, non-cancer surgeries in the Medicare population. The selective choice of four cardiovascular surgeries reminded me of the Baskervilles authors' careful selection of heart disease categories.

The authors justified their surgery selections by references to four studies that had investigated the relationship between surgical mortality and (1) surgeon age, (2) surgeon experience, (3) hospital volume, and (4) surgeon age and sex. Two of the co-authors of the fourth study were also co-authors of the birthday study.

I compared the birthday study with these four papers in order to see if the surgery choices may have been made to bolster the case the authors wanted to make. One paper considered six cardiovascular and eight cancer operations; two papers examined four cardiovascular and four cancer operations; and the fourth paper considered four cardiovascular surgeries and the sixteen most common non-cardiovascular surgeries in the Medicare population. The birthday paper's choice of seventeen surgeries is suspiciously peculiar.

None of the four referenced studies excluded patients with cancer but the birthday study did, with this unconvincing explanation: "To avoid patients' care preferences (including end-of-life care) affecting postoperative mortality."

In addition, the birthday study defined operative mortality as death within thirty days after surgery. All four of the referenced papers (including the paper with overlapping co-authors) defined operative mortality as death before hospital discharge or within thirty days after the operation. One paper explains that, "Because, for some procedures, a large proportion of operative deaths before discharge occurred more than 30 days after surgery, 30-day mortality alone would not adequately reflect the true operative mortality."

The *BMJ* publishes rapid responses to its articles and there were several reactions to the birthday paper, including a comment by me and these snippets from a long thoughtful piece:

The suggested scenarios sound implausible to this surgeon, but may seem plausible to the authors of this paper only one of whom has been to medical school, and none of whom have any surgical experience . . .

[T]his paper is driven by an hypothesis so odd that one wonders if it was . . . the result of a fishing expedition in search of statistical significance. It seems unlikely that the authors undertook the substantial task of cleaning this dataset solely to look at the effect of surgeon's birth date on mortality . . . How many other hypotheses did the authors examine, and reject, because they did not lead to a publishable p-value? We have no way of knowing for sure, but all the hallmarks of p-hacking are on display in this paper.

P-hacking . . . is, of course, statistical malpractice, but seems likely to have been the approach taken by these authors. Marginally significant p-values (0.03) such as the authors report after very significant statistical exertions (29 pages of Online Supplement) are typical of p-hacking . . . Moreover, not only are the p-values reported by the authors marginal, but the effect size is minuscule: . . . a total of 29 deaths . . . over 4 years and almost one million operations . . . spread over a total of 47,489 surgeons. . . .

The problems with this manuscript are so profound that Kato's willingness to impugn the reputation of every surgeon in America is not just wrong; it's irresponsible. Kato's manuscript's conclusions have already been picked up by the lay press, and are rapidly spreading over the internet as "click bait", no doubt boosting the authors' careers, but in the process undermining, patients' trust in their surgical health care team.

The Brady Bunch

The 2013 Christmas issue included a paper with an alluring title, "The Brady Bunch?," which was concerned with "nominative determinism," the idea that our surnames influence our choice of professions. With my name *Smith*, I might have been predestined to choose to be a blacksmith or silversmith. That didn't happen but a newspaper article did find "a dermatologist called Rash, a rheumatologist named Knee, and a psychiatrist named Couch." The authors of the *BMJ* paper add their own examples:

Was it a surprise that during the London 2012 Olympics, a man who shares his name with lightning became the fastest man in the world? Usain Bolt ran the 100 metre final in 9.63 seconds, faster than anyone else in history. It is also somewhat predictable that Bulgarian hurdler Vania Stambolova would, well, stumble over as she did, unfortunately falling at the first hurdle in her heat.

Not every dermatologist is named Rash nor is every runner named Bolt. Of all the many people in any profession, there are bound to be some whose names are coincidentally related to their profession. The question is whether this is true for an unusually large number of people.

The authors focused on whether Dubliners named Brady have an unusually large incidence of bradycardia. Now, unless you are a doctor, you are probably wondering (as I was) what bradycardia is. The word *bradycardia* comes from the Greek words *bradys* ("slow") and *kardia* ("heart"), and signifies a slower than normal heart rate which may be treated with the surgical implantation of a pacemaker.

The argument that people named Rash might be attracted to professions dealing with rashes assumes, reasonably enough, that they know what the word *rash* means. The fact that *bradycardia* is an unfamiliar word for the overwhelming majority of the population demolishes the idea that people named Brady would be psychologically susceptible to bradycardia.

Nonetheless, the authors looked at 999 pacemaker implants at an unnamed hospital in Dublin during the period January 2007 through February 2013 and found that 8 (0.80 percent) were implanted in patients named Brady. For comparison, they looked at 161,967 residential telephone listings in the area of the hospital and found that 579 (0.36 percent) were named Brady. The p-value for this disparity is 0.03, which is (marginally) statistically significant.

Given the utter implausibility of the hypothesis, how did the authors find a statistically significant pattern? There are several ways in which the data may have been p-hacked, starting with the fact that the authors may have looked at several diseases in addition to bradycardia.

Focusing on bradycardia, there are eight Dublin hospitals in which pacemaker implants are performed, and the authors are affiliated with three of them, yet results are only reported for a single hospital. Did they omit hospitals with contradictory data?

The authors say that the data were for a 61-month period from January 1, 2007, through February 28, 2013. This is actually a 74-month period, which raises questions about how carefully the calculations reported in the paper were carried out. Readers might also wonder why the time period is approximately 6 years, instead of the more obvious 5 or 10 years? Also, why did the time period end on February 28, 2013, rather than December 31, 2012, which would have been exactly six years, or end in a month closer to the December 2013 publication date? Did January and February 2013 help the case they were trying to make while March, April, and May 2013 hurt the case?

The frequency with which the surname Brady appears in telephone directories is a noisy measure of the number of Bradys who were candidates for pacemaker implants. For one thing, pacemaker recipients are somewhat older than the telephone directory population and perhaps the prevalence of the surname Brady has declined over time—and is consequently more common among older citizens. Second, even though the directory is labeled *residential*, some entries are not households. My search for *Ltd* and *Limited* in the Dublin residential directory turned up dozens of unique non-household results (some listings appear multiple times), including

Findings Ireland Ltd
TCOM Ltd
DecathlonSports Ireland Ltd
Roofing Ltd Main Street
Anbawn Computing Limited
Services Limited BROC Accounting

I also stumbled across these listings:

Balbriggan Library
National Council for the Blind
Flyover B&B
The Swedish Coffee Company

There is no easy way to determine how many of the residential Dublin listings are duplicates or not households, but it is clear that the 0.03 p-value is misleading and that the many issues with this paper make the statistical evidence unconvincing.

In addition, we need to consider "oomph," the practical significance of the findings. Is the difference between a 0.80 percent and 0.36 percent incidence of any real importance? Should people named Brady fear bradycardia? Should doctors test everyone named Brady for bradycardia? Or should we lament that this paper did not win an Ig Nobel award?

Motorcycles, Full Moons, and Gardens Full of Forking Paths

Fragile research can often be described by statistician Andrew Gelman's analogy to a "garden of forking paths." Imagine, as depicted in Figure 7.1, that a researcher is walking through a garden with a large number of forking paths. The chosen paths lead to a destination but it is not an inevitable destination. Different choices would have led to different destinations.

Sometimes, a researcher tries several paths and only reports one; this is *p-hacking*. Other times, a researcher tries just one path and makes up a theory after reaching the destination; this is *hypothesizing after the results are known—*or *HARKing*.

An example of the former is when an ESP researcher tests a thousand people and only reports the results for the most successful guesser. An example of the

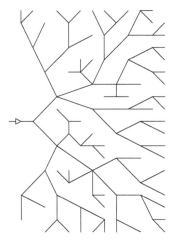

Figure 7.1 A garden of forking paths.

latter is when an ESP researcher analyzes the data after a test is finished and finds positive ESP at the beginning and end of the experiment and negative ESP in the middle. Often, as with most ESP studies, researchers don't realize that they have walked through a garden of forking paths.

An interesting example with many forking paths is a 2017 Christmas paper asserting that fatal motorcycle accidents are more likely when there is a full moon. The image of a full moon conjures up thoughts of craziness, supernatural activities, and werewolves. Indeed, the Roman moon goddess Luna is the source of the words *lunacy* and *lunatic*, and many nurses and police officers believe that there is an increase in wild and crazy activities on full-moon nights though there is little evidence supporting their anecdotal recollections.

Nonetheless, the image of motorcycle riders driving fast and furiously during full moons is seductive and this research, co-authored by prominent professors at the University of Toronto and Princeton University and published in the *British Medical Journal*, was reported far and wide by the media, including the *Washington Post*, *Guardian*, and *Psychology Today*.

It turns out that the authors were not arguing that full moons make people behave in crazy ways. Their argument is that drivers are distracted by full moons and pay less attention to possible driving hazards:

Glancing at the full moon takes the motorcyclist's gaze off the road, which could result in a loss of control . . . [W]e recommend riders and drivers orient their attention, ignore distractions and continuously monitor their dynamic surroundings.

This is a reasonable argument and a sound recommendation. Would anyone endorse being distracted while driving? However, an empirical test of this theory involves several forking paths, including which vehicles, how fatalities are counted, which hours, which nights, and which control group.

Which Vehicles

The *BMJ* paper used data from the National Highway Traffic Safety Administration's Fatality Analysis Reporting System (FARS) to count the number of accidents that involved motorcycles. However, motorcyclists are not the only drivers who might be tempted to stare at a full moon. In fact, drivers riding in closed cars and trucks may have to go through extensive contortions to get a clear view of the moon—which is an argument for considering all vehicle accidents, maybe even prioritizing vehicles with limited visibility.

In addition, many fatal motorcycle accidents involve collisions with other vehicles, and it may have been the driver of the other vehicle, not the cyclist, who was distracted. It is also possible that motorcycle headlights and taillights might not be as brightly visible on full-moon nights. On the other hand, it might be safer to drive on full-moon nights, when there is more road visibility. The situation is clearly not straightforward.

In addition to data on fatal accidents involving motorcycles, FARS has data for accidents involving mopeds, three-wheel motorcycles, off-road motorcycles, and other motored cycles. These might have been included with motorcycles. Other researchers might consider all fatal vehicle accidents or focused on, say, pickup trucks, convertibles, or SUVs. There is no inherently best choice, but there are several forking paths that might lead to different conclusions.

How Fatalities Are Counted

Another forking path is that the FARS database has data on the number of fatal accidents and the number of fatalities in each accident. In quantifying the risks from driving on full-moon nights, risk might be better gauged by the chances of someone being killed rather than the chances of being involved in an accident in which someone dies.

Either way, a calculation of risks should also take into account how many vehicles are on the road and either the aggregate time spent on the road or the total number of miles driven on full-moon nights. Unfortunately, the FARS system does not have these data; so we have to make do with either the number of fatal accidents or the number of accident fatalities.

Which Hours

Since the argument is that drivers are distracted by the sight of a bright full moon, the fatalities data should be restricted to hours when the moon is visible. There are many forks with alternative hourly windows. The *BMJ* paper used the 16-hour time period from 4 p.m. on the full-moon date to 8 a.m. the next morning. However, it is seldom dark at 4 p.m. or 8 a.m. and a full moon cannot be seen in the daytime. How could drivers be distracted by a full moon that they literally cannot see? If traffic accidents are being caused by a distracting full moon, the strongest evidence should be during those hours when the sky is darkest and the full moon is most visible. One reasonable alternative is the 8-hour window between 8 p.m. to 4 a.m.

Which Nights

It might seem relatively straightforward to identify nights with a full moon; however, there are many forking paths. The *BMJ* paper obtained full-moon dates from timeanddate.com:

We defined the full moon as the one night each month when the entire facing surface was illuminated, as viewed from Earth. On rare occasions, a month had two full moons; in such cases we included both appearances.

The authors of the *BMJ* paper did not, nor did I initially, recognize that full-moon dates depend on the time zone from which the moon is observed. The continental U.S. has four time zones with a three-hour difference between the Eastern and Pacific zones. The Alaskan time zone is one hour behind Pacific time and the Hawaii time zone is two more hours behind Alaska time. The FARS fatality data also include Puerto Rico and Virgin Islands, which are both in the Atlantic Time Zone, one hour ahead of the Eastern Time Zone.

A full moon can occur on one date in one time zone and on a different date in another time zone. For example, the June 1998 full moon occurred on June 10 in the Eastern time zone and on June 9 in the five other U.S. time zones. Further complicating matters, thirteen U.S. states span two time zones and some U.S. counties are also split into two time zones.

A completely accurate identification of whether a fatal accident occurred on a full-moon date at the geographic location of the accident would be extremely tedious and sometimes impossible, because the FARS data on the city (or nearest city) are very spotty.

It is understandable that the authors of the *BMJ* paper used a single time zone to identify the dates of full-moon nights; however, they inadvertently chose the London time zone, which is five hours ahead of the U.S. Eastern Time Zone and 8 hours ahead of the Pacific Time Zone. Early morning full moons in England are often full-moon evenings the day before in the United States. For example, the full moon that occurred at 1:42 a.m. on March 21, 2019, in London occurred on the evening of March 20 in the United States (at 6:42 p.m. in Los Angeles and 9:42 p.m. in New York).

A related issue is that, after identifying a full-moon date, the *BMJ* paper recorded fatal accidents that occurred between 4 p.m. that evening and 8 a.m. the next day. Consider, however, the full moon that occurred at 12:05 a.m. on October 20, 1975, in the U.S. Central Time Zone. Instead of counting accidents that occurred the night of October 20, it seems more in keeping with the spirit of the paper to have the full-moon window include the time of the full moon—specifically, accidents that occurred the night of October 19.

Among the many possible paths, one reasonable alternative is the Central Time Zone, which is heavily populated and roughly in the middle of the continental United States, and a nighttime window that includes the time of the full moon.

Which Control Group?

The *BMJ* paper compared the number of accidents on full-moon days to the average number of accidents seven days before and seven days after the full-moon day. That is certainly a plausible path, but there are others. One natural alternative is fourteen days before and fourteen after, which would be (approximately) the dates of new moons, which are essentially the opposite of full moons. The seven- and fourteen-day control groups both have the advantage of comparing weekdays to weekdays and weekends to weekends.

Results

There were a large number of forking paths in this research and I didn't look at all of them. In addition to motorcycles, I looked at all fatal vehicle accidents. In addition to the number of accidents, I looked at the number of fatalities. In addition to a plus-or-minus-seven-day control group, I used plus-or-minus fourteen days. In addition to a 4 p.m. to 8 a.m. window, I used the time interval 8 p.m. to 4 a.m. Instead of the London time zone with a window beginning on the evening of the full-moon date, I used the U.S. Central Time Zone with a window including the time of the full moon. The results of these strolls through the garden are shown in Table 7.1.

There is no compelling best route through this garden of forking paths—which is precisely the point. When there is no best path, we should consider a variety of plausible choices and see whether the conclusions depend on the choices.

Here, the *BMJ* path (the bolded entries in Table 7.1) gives 2.9 percent more fatal motorcycle accidents on full-moon nights than on control nights—3706 versus 3603. (Using the London time zone, the *BMJ* paper reported a 5.3 percent difference.) This conclusion is not robust since other reasonable paths do not indicate that it is particularly dangerous to drive on full-moon nights. Of the sixteen comparisons in Table 7.1, the number of accidents or fatalities were higher on full-moon nights in eight cases and lower in eight cases. For fourteen of the sixteen paths, there is less than a 1 percent difference between full-moon nights and the control group.

Twenty years ago, the time it would have taken to follow alternative paths through large databases was daunting, which often dissuaded researchers from investigating their gardens widely or deeply. Today, powerful computer algorithms can be used to explore many paths. This capability should not be abused by *p*-hacking or HARKing, but can be used to test the robustness of a study's conclusions. Here, the evidence that it is unusually dangerous to drive on full-moon nights is not compelling.

Table 7.1 *Number of Accidents and Fatalities on Full-Moon Nights*

	4 p.m. to 8 a.m. window		8 p.m. to 4 a.m. window	
	All Vehicles	Motorcycles	All Vehicles	Motorcycles
Number of fatal accidents				
Full moon	37,964	**3,706**	21,158	1,985
+/- 7 days	38,075.5	**3,603.0**	20,979.5	1,974.5
+/- 14 days	38,047.0	3,669.5	21,272.0	1,997.0
Number of fatalities				
Full moon	42,222	3,873	23,637	2,080
+/- 7 days	42,414.5	3,768.0	23,433.0	2,073.5
+/- 14 days	42,462.5	3,828.5	23,834.0	2,089.5

Note: Control nights are an average of the accidents that occurred either 7 or 14 days before and after the full-moon date, using the same time window.

Too Much to Read

The Bible (Ecclesiastes 12:12, King James Version) warns that "Of making many books there is no end; and much study is a weariness of the flesh." A 2010 Christmas paper reported an empirical update of this biblical admonition. The authors searched the database of the U.S. National Library of Medicine for papers that referred to echocardiography and found 113,976 papers—which is a long reading list for a newly minted doctor specializing in echocardiography:

We assumed that he or she could read five papers an hour (one every 10 minutes, followed by a break of 10 minutes) for eight hours a day, five days a week, and 50 weeks a year; this gives a capacity of 10000 papers in one year. Reading all papers referring to echocardiography ... would take 11 years and 124 days, by which time at least 82142 more papers would have been added, accounting for another eight years and 78 days. Before our recruit could catch up and start to read new manuscripts published the same day, he or she would—if still alive and even remotely interested—have read 408,049 papers and devoted (or served a sentence of) 40 years and 295 days. On the positive side, our recruit would finish just in time to retire.

Bear in mind that this calculation was done more than a decade ago and the backlog of papers for a new doctor's reading list is much larger now than it was then.

The authors recognized that papers are not all of equal quality but that separating the wheat from the chaff is difficult because worthwhile papers are not confined to a small set of journals.

A more promising strategy is to ignore all papers that do not use randomized controlled trials. Surprisingly, only 457 of the 113,976 echocardiography papers did! The authors' conclusion might well be interpreted as a manifesto that almost all echocardiograph research is not worth reading:

In fields such as diagnostic imaging the plethora of publications hides the fact that we still lack sufficient evidence for rational practice . . . . The best way to assess any diagnostic strategy is a controlled trial in which investigators randomise patients to diagnostic strategies and measure mortality, morbidity, and quality of life, but only 2.4% of diagnostic recommendations in the guidelines from the American Heart Association and the American College of Cardiology are supported by this level of evidence.

Retroactive Prayer

I was intrigued by a 2001 Christmas paper about how remote prayer by strangers had reduced the duration of fever and hospital stays of patients who had been hospitalized for bloodstream infections. The author, Leonard Leibovic, considered all adult patients at an Israeli university hospital during the years 1990 through 1996 who had bloodstream infections. The patients were randomly divided into an intervention group that received a brief remote prayer for good health and full recovery. Those in the control group were ignored (though they may well have been prayed for by friends and relatives).

Mortality was slightly lower for the prayer group (28.1 versus 30.2 percent), though the difference was not statistically significant. On the other hand, there were statistically significant reductions in the duration of fever and hospital stays for those in the prayer group.

This conclusion certainly seemed ripe for further investigation. One thing I noticed right away was that the data were skewed by outliers. For example, the median hospital stays were very similar (seven days for the treatment group and eight days for the control group) but the maximum stays were extremely large and very different (165 days for the treatment group and 320 days for the control group). I also realized that the length of a hospital stay is probably closely related to the duration of the fever and that, paradoxically, a short fever and/or hospital stay is not necessarily a good thing since both can end, not with recovery, but death. This reminded me of a study that found that sepsis patients who had very low blood pH levels when they were discharged from a hospital were unlikely to be readmitted—not because they were fully recovered but because they had been discharged to the mortuary.

When I read the prayer paper more carefully, I discovered this surprising description of the treatment:

As we cannot assume a priori that time is linear, as we perceive it, or that God is limited by a linear time, as we are, the intervention was carried out 4-10 years after the patients' infection and hospitalization.

I read that sentence twice and realized that this was a prank paper. The prayers were said in the year 2000 for patients who had been hospitalized from 1990 through 1996, all of whom had already either died or recovered. The author evidently intended to remind us that randomized controlled trials sometimes have false positives.

Not everyone got the joke. Many cited the paper as proof of the power of prayer. A prominent psychic researcher proclaimed that the study was among "the largest, most important, and best-funded research studying consciousness and nonlocality." Some of those who got the joke objected to prank papers like this precisely because some people were tricked into taking them seriously. In a prank response to this prank paper, one hospital manager objected that "it is morally unacceptable to intervene experimentally in the routine care of a patient without his or her permission."

The Irony

Fame, fortune, and funding await those who publish memorable research. Unfortunately, the lure of these rewards leads many to p-hack their way to semi-plausible, provocative results that will be reported widely. The achievements of scientists created the reward system; the scientific demand for rigor created the statistical significance requirement; and the scientific journals that disseminate research created the p-value launchpad for announcing eye-catching findings. The unintended consequence is that, even the best journals now publish rubbish that makes some question the sanity of scientists and the value of scientific research.

PART III

Data Mining

CHAPTER 8

Looking for Needles in Haystacks

It is tempting to avoid the hard work of thinking about sensible ways of predicting something and, instead, simply let a computer try many different possibilities and choose the variables that make the most accurate predictions. Letting a computer select the best model is called *data mining* (aka *machine learning* or *knowledge discovery*). Alas, the easy way isn't always the best way.

Sometimes, data mining involves Hypothesizing After the Results are Known (HARKing)—after the computer finds a statistical relationship, think of reasons, any reasons. Other times, data miners argue that no reasons are needed. All that matters is that it works. Do we need to know how our smartphones work? Do we need to know why one variable predicts another? The problem is that if there is no underlying reason, then the discovered statistical relationship is just a meaningless coincidence. Noticing a coin landing heads three times in a row won't help us predict the next coin flip.

In addition to avoiding hard thinking, data-mining enthusiasts say that they might uncover heretofore unknown relationships. Writing in the *Harvard Business Review*, the Vice President of Thought Leadership (what a great job title!) at software giant SAS argued that "Traditional database inquiry requires some level of hypothesis, but mining big data reveals relationships and patterns that we didn't even know to look for." Similarly, an article in the respected technology magazine *Wired* was titled "Big Data and the Death of the Theorist," with the author explaining that "The algorithms find the patterns and the hypothesis follows from the data. The analyst doesn't even have to bother proposing a hypothesis any more."

Decades ago, data mining was considered a sin comparable to plagiarism because, as with p-hacking, finding models that predict the past is easy and useless. Today, the data-mining plague is seemingly everywhere, cropping up in medicine, economics, management, and even history.

For example, a computer might be given the daily stock prices of a Chinese tea company over a two-year period and a list of the words Donald Trump tweeted each day over that same period. The computer program can calculate thousands of correlations between how often each word was tweeted and the tea company's stock price one to five days later and might determine that the highest correlation was between Trump tweeting the word *with* and the tea company's stock price four days later.

The problem is that if we look at thousands of correlations, even correlations among random variables, some are bound, by luck alone, to be statistically significant. Remember that, on average, one out of every twenty correlations among random numbers will be a false positive. With a thousand correlations, we expect fifty false positives.

This was less of a problem when it was not practical to calculate thousands of correlations. Now, computers can do in seconds what it would take humans years to do by hand. Plus, computers do not get tired or make arithmetic mistakes—computers are relentless data miners.

On the other hand, computers have no way of determining whether the statistical relationships they discover make sense. In our Trump-tweet example, computers do not know what *stock price*, *Chinese tea company*, or *days* means. Computers do not know what a *Trump* or a *tweet* is or the meaning of any of the words Trump tweeted. The labels might as well have been randomly chosen car license plates. Computers have no way of assessing whether the correlation between Trump tweeting *with* and the tea company's stock price is a plausible model for predicting the stock price. If the computer program's task is to buy and sell stocks, it might buy and sell this stock based on how many times Trump tweets the word *with*, and lose a lot of money doing so.

Humans know better. We have the ability to say, "Nonsense!" We just need to recognize that most of the patterns discovered by data-mining expeditions are nonsense.

Mining Fool's Gold

One professional data miner gushed that

Twitter is a goldmine of data . . . . [T]erabytes of data, combined together with complex mathematical models and boisterous computing power, can create insights human beings aren't capable of producing. The value that big data Analytics provides to a business is intangible and surpassing human capabilities each and every day.

Notice how easily he assumes that large doses of data, math, and computing power make computers smarter than humans.

He cites a study of the Brazil 2014 presidential election as an example of how data mining Twitter data "can provide vast insights into the public opinion." The author of the Brazilian study wrote that

Given this enormous volume of social media data, analysts have come to recognize Twitter as a virtual treasure trove of information for data mining, social network analysis, and information for sensing public opinion trends and groundswells of support for (or opposition to) various political and social initiatives.

Brazil's 2014 presidential election was held on October 5, 2014, with Dilma Rousseff receiving 41.6 percent of the vote and Aécio Neves 33.6 percent. As no one received more than 50 percent, a runoff election was held three weeks later, on October 26, which Rousseff won 51.6 to 48.4 percent.

To demonstrate the power of data mining social media, the computer algorithm mined the words used in the top ten Twitter Trend Topics in 14 large Brazilian cities on October 24, 25, and 26 in order to predict the winner in each city. It sounds superficially plausible, but there are plenty of reasons for caution. Those who tweet the most are not necessarily a random sample of voters. Many are not even voters. Many are not even people. Remember the discussion in earlier chapters about how bots and other trickery are used to command the trend. Instead of reflecting the opinions of voters, the Twitter data may mainly reflect the attempts of some to manipulate voters.

How well did the algorithm do? The study proudly reported a 60 percent accuracy rate. That's somewhat better than flipping a coin but seemed odd to me, living in the United States, where most big cities reliably vote for one party or the other. It is not at all difficult to predict how Boston, Chicago, and New York will vote in presidential elections. Here, comparing the results in the first and second rounds of the 2014 Brazil presidential election in these fourteen cities, if one had simply predicted that whoever did better in the October 5 vote would win the October 26 runoff, the accuracy would have been 93 percent.

Despite the superficial appeal of using super-fast computers to mine mountains of data, this model's discoveries were fool's gold.

Trump's Tweets

In 2022 Twitter was reported to have more than 200 million unique daily users sending a total of 500 million tweets a day, which adds up to nearly 200 billion tweets a year. What incredible opportunities these vast data offer for finding useless statistical patterns!

Many have done just that, including some JP Morgan researchers who found statistically significant correlations between interest rates and the number of Donald Trump tweets containing the words *China, billion, products, Democrats,* and *great.*

Such correlations are trivial to find and of trivial usefulness. I once demonstrated this with a data-mining exploration of Trump's tweets during the 3-year period beginning on November 9, 2016, the day after his election as President of the United States. I found a 0.43 correlation between Trump tweeting the word *president* and the S&P 500 index of stock prices two days later. Specifically, the S&P 500 was predicted to be 97 points higher two days after a one-standard-deviation increase in Trump tweeting the word *president.* The p-value was essentially zero—and so was the usefulness of this correlation.

I did not choose the word *president* in advance. I considered thousands of tweeted words, the Dow Jones Average as well as the S&P 500, and predictions one to five days ahead. I was sure to find some statistically significant coincidental correlations and I did.

To make the point even more strongly, I did more data mining with even more fanciful variables. I found that the low temperature in Moscow is predicted to be 3.30°F higher four days after a one-standard-deviation increase in Trump's use of the word *ever,* and the low temperature in Pyongyang is predicted to be 4.65°F lower five days after a one-standard-deviation increase in the use of the word *wall.*

I even considered the proverbial price of tea in China. (Yes, the hypothetical example given earlier in this chapter was not hypothetical.) I couldn't find daily data on tea prices in China, so I used the daily stock prices of Urban Tea, a Chinese tea product distributer, and found that Urban Tea's stock price is predicted to fall four days after Trump uses the word *with* more frequently.

As if that were not absurd enough, I generated a random variable and used my data-mining program to find some Trump-tweeted words that were correlated with these random numbers. I was not disappointed. A one-standard deviation increase in Trump's use of the word *democrat* had a strong positive correlation with these random numbers five days later.

These increasingly implausible statistical patterns are intended to demonstrate that data-mining expeditions can generally be counted on to find statistically significant patterns that can generally be counted on to be coincidental.

I wrote a satirical piece about these data-mined Trump tweets that concluded, tongue firmly in cheek,

I did not anticipate what I uncovered—which is compelling evidence of the value of using data-mining algorithms to discover statistically persuasive, heretofore unknown correlations that can be used to make trustworthy predictions.

I naively assumed that readers would get the point of this nerd joke—data-mining can easily discover patterns that are utterly useless. I submitted the paper to an academic journal and one of the reviewer's comments demonstrates beautifully how deeply embedded is the notion that statistical significance supersedes common sense:

The paper is generally well written and structured. This is an interesting study and the authors have collected unique datasets using cutting edge methodology.

The second reviewer got the joke but did not appreciate my criticism of data mining:

There are already documented cases where data-mining has been used to make practical and useful predictions. The author's findings and viewpoints are therefore not value adding.

Watson, Our Computer Overlord

When IBM's Watson computer system defeated Jeopardy champions Ken Jennings and Brad Rutter in 2011, the world took notice. Watson seemed to understand the questions as well as the humans and to have an encyclopedic knowledge that surpassed theirs. The black box perched between Jenner and Rutter and the synthesized voice Watson used to answer questions added to the illusion that the computer was human-like, but much smarter. As Watson's victory became inevitable, Jennings wrote on his video screen, "I, for one, welcome our new computer overlords." Rutter seemed less welcoming. The reality is that an IBM research team had spent several years working on one specific task—building a computer that could win at Jeopardy. They produced an extremely powerful data-mining machine, not anything that could be remotely considered intelligent.

The question-processing part of the program was perhaps the most impressive achievement—parsing the question into components that could be

used to determine the best answer. For example, the key components of this hypothetical question

This pilot went into outer space in 1961

are *this*, *pilot*, *outer space*, and *1961*.

Even though a computer program does not know what a pilot is, it can search through a very long list of labels that had been grouped into the categories Person, Place, or Thing and find that the label *pilot* is in the Person category.

The label *this* before the person label *pilot* meant that the correct answer is a person's name and, since Jeopardy answers must be in the form of a question, the answer should be in the form "Who is . . ." (or "Who was . . ." if the person is deceased). The three labels *pilot*, *outer space*, and *1961* tell the computer that it should do a data-mining search for a person-name label that is linked to these three labels.

The understanding of the question is trivial for humans because we know what every word in the question means, but the identification of relevant labels is enormously challenging for a computer algorithm processing text it does not understand. Watson used hundreds of different algorithms running simultaneously to parse the question in different ways and to data-mine its database in search of matches.

The relatively easy part was finding answers that had the matching identifier labels, particularly since more than 95 percent of Jeopardy answers are the titles of Wikipedia pages. Watson's enormous database includes a compilation of everything in Wikipedia (and some other references too). Watson's Person category included every name that appears in Wikipedia, each linked to identifier labels.

The more agreement there was among the algorithms, the higher the probability that the most popular answer was correct. In this hypothetical example, a person's name that has the identifier labels *pilot, outer space*, and *1961* is Yuri Gagarin, which is the correct answer (and the title of a Wikipedia page).

Watson's secret advantage was that its electronic button-pushing finger was faster than the human fingers that Jennings and Rutter were hobbled with. The illusion was that Watson figured out answers faster; the truth was often that Watson pushed its button faster.

The questions that gave Watson trouble were those that it had difficulty parsing into useful identifiers. One memorable flub was this Final Jeopardy clue in the category U.S. Cities:

Its largest airport is named for a World War II hero;
its second largest for a World War II battle.

The correct answer is Chicago, but Watson answered Toronto, which is not a U.S. city.

The manager of the Watson project explained that the team had learned that category names are often misleading. The category label "U.S. City" does not mean that the answer is the name of a city. The question might have been, "The windy city is on this body of water," with "What is Lake Michigan?" being the correct answer. Because of such ambiguities, Watson largely ignored category names. If "U.S. City" had been part of the question, Watson would have honed in on that. To the extent that Watson did consider "U.S. Cities," it might have been confused because some web pages identify Toronto as an "American League City," referring to its baseball team and the labels America and U.S. are often used interchangeably.

In this particular Final Jeopardy question, the humans recognized that the correct answer had to be a U.S. city with two airports. New York and Chicago were obvious possibilities and New York's JFK Airport is named after a World War II hero. LaGuardia Airport was named after former mayor Fiorello La Guardia, who was not a World War II battle; so Jenner and Rutter could reason that Chicago was the likely answer, even if they didn't know that Chicago's

O'Hare Airport was named after World War II hero Edward "Butch" O'Hare and that the Midway Airport was named after the World War II Battle of Midway.

Watson tackled the question differently. Parsing the first part of the sentence, the label *its largest airport* suggested a place and the identifiers *largest airport* and W*orld War II hero* suggested New York, Chicago, and Toronto, whose Pearson International Airport is named after former Prime Minister Lester B. Pearson, who had served in World War II in several capacities, including carrying secret documents to Europe.

Watson stumbled on the second part of the question because it did not know what the word *its* in "its second largest" referred to. This is an example of a more general computer hurdle known as the Winograd schema challenge. What does *it* refer to in this sentence?

The trophy doesn't fit into the brown suitcase because **it** is too large.

This question is trivial for humans because we know what trophies and suitcases are and what *doesn't fit* and *too large* mean, so we know that *it* refers to the trophy. Since computers don't know what words mean, they find Winograd schemas challenging.

Here, Watson did not know that *its* in "its second largest" referred to the city alluded to in the first part of the question. Adding to Watson's confusion, "named for" does not reappear in the second part of the question.

Based on matches it found, Watson's best guess was Toronto, though it estimated that there was only a 14 percent chance that this was the correct answer.

Dr. Watson

In the popular media, robots are generally smarter—much smarter—than humans, including doctors and nurses. In the movie *Big Hero 6*, for example, Baymax is a lovable super hero and all-knowing healthcare provider that has been programmed to have the feelings and emotions of its creator's father. In real life, instead of Baymax, we have IBM's Watson, which has no feelings or emotions and is little more than a sometimes-helpful record keeper.

Soon after Watson defeated the best human Jeopardy players, IBM boasted that Watson would revolutionize healthcare. An IBM Senior Vice President for Cognitive Solutions and Research boasted that "Watson can read all of the healthcare texts in the world in seconds, and that's our first priority, creating a 'Dr. Watson,' if you will." Yes, Watson can store more data and retrieve data faster than any doctor. The unwarranted assumption is that data-mining algorithms can come up with the best diagnoses and treatments and even discover new diagnoses and treatments.

There are good reasons for skepticism. The practice of medicine is much more than storing, retrieving, and mining data. Computer algorithms have no effective way of separating the best research papers from the thousands of so-so papers and garbage. Nor do they have any reliable way of recognizing when previously reported results have been reversed by subsequent studies.

Computer algorithms also struggle with complex medical conditions involving multiple health problems which are the norm in the real world, particularly with elderly patients. Dr. Watson would not know the practical relevance of a retrieved study of the benefits and side effects of one medication on one disease.

So, how has it worked out? Have computer data-mining algorithms replaced doctors? Shortly after Watson's Jeopardy win, the MD Anderson Cancer Center in Houston began employing Watson, accompanied by great hope and hype. A story headlined "IBM supercomputer goes to work at MD Anderson" began

First he won on Jeopardy!, now he's going to try to beat leukemia. The University of Texas MD Anderson Cancer Center announced Friday that it will deploy Watson, IBM's famed cognitive computing system, to help eradicate cancer.

The idea was that Watson would data-mine enormous amounts of patient data and research papers, looking for clues to help it diagnose and recommend treatments for cancer patients based on the patterns it discovered. Cancer would be extinguished!

Watson flopped because winning at Jeopardy and eradicating cancer are very different tasks. Five years and $60 million later, MD Anderson fired Watson after "multiple examples of unsafe and incorrect treatment recommendations." Internal IBM documents recounted the blunt comments of a doctor at Jupiter Hospital in Florida: "This product is a piece of s—. . . . We can't use it for most cases."

IBM spent more than $15 billion on Dr. Watson with no peer-reviewed evidence that it improved patient health outcomes. Watson Health disappointed so soundly that, after more than a year spent looking for buyers, IBM sold the data and some of its algorithms to a private investment company in January 2022 for roughly $1 billion.

When Watson's foray into healthcare first started, IBM's CEO predicted that "Our moon shot will be the impact we have on health care. I'm absolutely positive about it." Watson ended up mainly being used for routine back-office tasks like record keeping, accounting, and billing. Shortly before Watson was sold for parts, an IBM television commercial boasted that Watson helps companies "automate the little things so they can focus on the next big thing." Talk about diminished expectations!

Dr. Watson was no Baymax.

The Irony

Scientists created the enormous databases and powerful computers that make it so easy to be seduced by the data-mining siren call: *Relax. The computer will find the best model.*

The erosion of science's credibility is aided and abetted by data mining because a virtually unlimited number of coincidental sequences, correlations, and clusters can be discovered in large databases. Those who use data-mining algorithms to discover something—anything—are sure to succeed. Those who do not appreciate the inevitability of patterns are sure to be impressed. Those who are seduced by these shiny patterns are likely to be disappointed. Those who act on the basis of data-mined conclusions are likely to lose faith.

CHAPTER 9

Beat the Market

Let's push an imaginary rewind button on the stock market and go back to April 14, 2015. The S&P 500 index of stock prices is at 2096, up 80 percent over the previous decade, which works out to be a 5.6 percent annual price increase. Add in dividends and investors enjoyed an average annual return above 7 percent.

Now, suppose the market were to go up a steady 6.3 percent over the next five years, with the S&P 500 beginning (as it did) at 2096 on April 14, 2015, and finishing (as it did) at 2842 on April 14, 2020. Add in dividends, and stockholders would have enjoyed a satisfying annual return close to 8 percent. Most investors would have been pleased by this serenely profitable history. They certainly would not have fled the stock market, vowing never to return.

This alternative history is not at all what happened. Figure 9.1 shows that the market surged more than 60 percent and then fell a breathtaking 34 percent before rebounding more than 25 percent. Instead of a peaceful stroll, it was a nauseating roller coaster ride. Many investors panicked after the 34 percent price collapse, sold everything, and vowed to give up on stocks—at least until the market returned to "normal"—as if there ever is a normal time in the stock market.

An investor who sells a stock after its price falls from $100 to $50 and plans to buy the stock back after the price returns to $100 is locking in a 50 percent loss. This behavior makes little or no sense but human psychology isn't always rational. I know a person who claims to have suffered post-traumatic stress disorder (PTSD) watching stock prices fall. He put everything in a checking

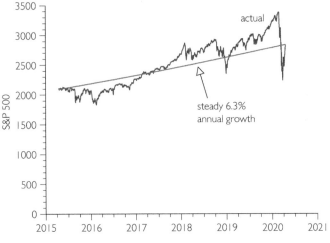

Figure 9.1 Ride the roller coaster.

account, earning nothing at all, with no intention of ever buying stocks again.

I also know financial advisers who refuse phone calls after a market dip because they want to protect their clients from the consequences of their own hysteria. It's not just naive retail investors. After the 23 percent market crash on October 19, 1987, the manager of a large college endowment made up excuses—I'm traveling to Europe, I'm traveling to Asia—to avoid meeting with the university's board. He knew that they wanted him to liquidate the university's stock portfolio. His evasiveness was rewarded as the stock market soon rebounded from its losses.

The fear that if stock prices fall, they will continue falling is not justified by history but by panic. If anything, the historical evidence is that large price movements in either direction—up or down—are generally due to investor overreaction to good or bad news and tend to be followed by price reversals.

When I managed a club soccer team, I used to say that the wins and losses are seldom as important as they seem at the time. The same is true of stocks. The latest zig or zag in prices is seldom as important as it seems at the time and has little or nothing to do with the long-run returns from stocks.

The average stock investor makes money in the long run because companies earn profits and pay dividends. In addition, if the economy, corporate profits, and dividends are substantially larger ten, twenty, and thirty years from now, stocks will be worth more then than they are now. If the economy does not grow for ten, twenty, and thirty years, then investors will have a lot more to worry about than their stock portfolios.

Bitcoin Babble

The crazy gyrations in bitcoin prices are additional evidence that investors aren't always rational. Since bitcoins generate no income, their intrinsic value is zero, yet people have paid hundreds, thousands, and tens of thousands of dollars for bitcoins.

In July 2021, Goldman Sachs' economics research team tried to explain this madness by arguing that, although bitcoin and other cryptocurrencies don't generate cash the ways stocks and bonds do, their fundamental value can be estimated from their "network size." Looking at eight cryptocurrencies, they found a "clear correlation" between the number of users and market value. Currencies with more users had higher market values.

When I read this, I muttered "data mining" and remembered the dot-com bubble back in the late 1990s. The Internet was new and exciting and companies that claimed to have something to do with the Internet were hot. The NASDAQ index, which was full of dot-com stocks, more than quintupled between 1995 and 2000. Companies could double the price of their stock simply by adding *.com*, *.net*, or *internet* to their names.

There was no conventional way to justify the soaring prices of dot-com stocks since few were profitable. So, creative analysts thought up new metrics for the so-called "new economy." One was *eye-balls*, the number of people who visited a company's web page; another was the number of people who stayed for at least three minutes. Even more fanciful was the number of *hits*—a web page's request for files from a server. Companies put dozens of images on a page and counted each image that was loaded from the server as a hit. Incredibly, analysts thought this meant something important.

Eventually the stock market came to its collective senses. Companies that don't make profits aren't worth billions of dollars. From its peak on March 10, 2000, the NASDAQ fell by 75 percent over the next three years.

In the same way, Goldman Sachs' bitcoin metric is preposterous. The number of people who own a cryptocurrency cannot remotely be considered a substitute for corporate profits and dividends. That's like saying that the value of Apple stock depends on how many shareholders it has, not how much profits the company makes. As I write this, Reliance Power, an Indian power-generating company, has approximately 4 million shareholders, more than any other company, but its market value is less than $50 billion. Apple's market value is $3 trillion.

Yet, Goldman Sachs, one of the world's most-respected investment banks, put its name on this fallacious nonsense. That's what comes from eyes-closed data-mining expeditions.

Benford's Law

As mentioned in this book's Introduction, the evidence is strong that bitcoin prices have been manipulated by pump-and-dump schemes in which unscrupulous players circulate boisterous rumors while they trade bitcoin back and forth among themselves at higher and higher prices, and then sell to the naive who are lured into the market by the get-rich-quick rumors and seemingly ever-rising prices.

One fascinating test of the pump-and dump theory is based on a remarkable relationship known as Benford's law. Suppose that we look at the annual profits of thousands of companies and record just the first (or *leading*) digit of each profit number; for example, *34,528* would be recorded as *3* and *422* would be recorded as *4*. We might think, reasonably enough, that each digit, *1* to *9*, would appear about the same number of times. Benford's law, however, predicts that the leading digit is more likely to be *1* than *2*, more likely to be *2* than *3*, and so on, as shown in Table 9.1. The leading digit is 6.5 times more likely to be *1* than *9*.

Table 9.1 *Benford's Law*	
Leading Digit	Probability
1	.301
2	.176
3	.125
4	.097
5	.079
6	.067
7	.058
8	.051
9	.046

This surprising relationship was discovered in the 1880s by a mathematics professor named Simon Newcomb. Before computers and the Internet, people looked up logarithms in books containing table after table of logarithmic values. Newcomb noticed that the early pages were dirtier and more worn than the later pages, leading him to speculate that numbers that begin with *1* are more common than other digits.

In the 1930s a physicist named Frank Benford confirmed Newcomb's speculative theory by tabulating the leading digits of all sorts of data (including the area of rivers and the numbers that appeared on the front pages of newspapers). He found that *1* was indeed the most common number, with the frequencies declining roughly as shown in Table 9.1. He called this puzzle the "Law of Anomalous Numbers." Soon there were mathematical proofs of situations in which the law (now called Benford's law or the Newcomb–Benford law) holds.

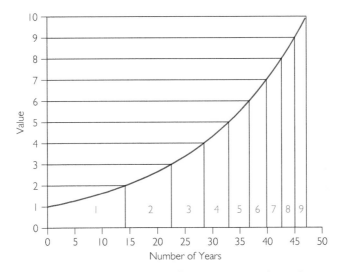

Figure 9.2 Benford's law with 5 percent compound growth.

For example, the distribution of leading digit of the *product* of randomly chosen numbers follows Benford's law. Suppose that a stock price or some other value starts at 1 and grows by 5 percent a year. The initial leading digit is 1, as shown by the red digit 1 in Figure 9.2. It takes 14.2 years for the value to reach 2, at which point the leading digit changes from *1* to *2*. The value reaches 3 and the leading digit changes from *2* to *3* in year 22.5; the value reaches 4 and the leading digit changes to *4* in year 28.4; and so on. Thus, it takes 14.2 years to go from 1 to 2 but only an additional 8.3 years to go from 2 to 3 and an additional 5.9 years to go from 3 to 4.

If we pick a date at random, the leading digit is more likely to be *1* than *2*, and more likely to be *2* than *3*, and so on. The probabilities are proportional to the distances between the vertical lines hitting the horizontal axis and are exactly equal to the probabilities shown in Table 9.1.

The underlying logic is that it takes longer to go from 1 to 2 (a 100 percent change) than to go from 2 to 3 (only a 50 percent change), so the value is more likely to be between 1 and 2 than between 2 and 3. The same argument applies if we start with a value of 10 and compare the intervals 10–20 with 20–30, or start with a value of 100 and compare 100–200 with 200–300, and so on. Benford's law does not require the percentage changes to be constant, only that the sequence of numbers is determined by percentage changes.

Benford's law doesn't apply if we are summing numbers instead of multiplying them. If we start with 10 and add 2 repeatedly, it takes just as long to go from 10 to 20 as it does to go from 20 to 30. Figure 9.3 shows a linear relationship that takes just as long to get from 1 to 10 as does the compound growth model in Figure 9.2. In contrast to Figure 9.2, a date picked at random is equally likely to have any of the nine possible leading digits.

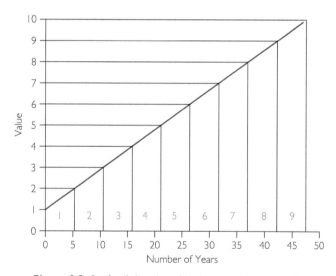

Figure 9.3 Benford's law doesn't hold with additive growth.

Benford's law should apply to stock prices because they rise and fall at compound rates. There are two ways to confirm this. First, looking at the prices of a large group of stocks at any point in time, approximately 30.1 percent of the leading digits should be *1*. Second, looking at prices over time for any single stock that has appreciated considerably, approximately 30.1 percent of the leading digits should be *1*.

To test this theory, I looked at the prices of all stocks traded on the New York Stock Exchange (NYSE) on March 7, 2022. The leading digit was *1* for 30.3 percent of the stocks and was *9* for 4.0 percent of the stocks, which are close to the 30.1 percent and 4.6 percent values implied by Benford's law. Figure 9.4 shows the full theoretical and empirical distributions. The close correspondence is striking for such a relatively small data set.

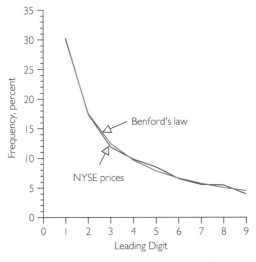

Figure 9.4 NYSE stock prices and Benford's law.

For a second test, Figure 9.5 compares the distribution of the leading digits of the price of Berkshire Hathaway stock over time. Again, the fit is not perfect but reasonably close.

Benford's law is more than just a surprising curiosity; it can also be used to detect fraud. When people make up data, they might be tempted to assume that, to avoid detection, the numbers should "look" random with every digit showing up equally often. However, if it is a situation where Benford's law ought to hold, there should be more leading-digit *1*s than *2*s, more leading-digit *2*s than *3*s, and so on.

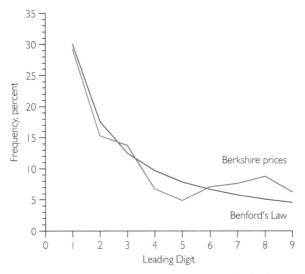

Figure 9.5 Berkshire Hathaway stock prices and Benford's law.

Since a company's revenue, costs, profits, and other measures of financial performance are likely to grow over time by percentages, they should be governed by Benford's law. Thus, Benford's law can and has been used to identify accounting fraud.

It might also be used to detect fraudulent bitcoin trading activity. Since increases in bitcoin prices are likely to be multiplicative, they should be governed by Benford's law. They are not. Figure 9.6 show the distribution of the leading digits of daily bitcoin prices back to 2014. Small digits appeared somewhat more often than large digits, but not nearly as often as implied by Benford's law.

There are 2,488 daily bitcoin prices underlying Figure 9.6, starting on September 17, 2014, with an initial price of $457.33. The daily percentage price changes over this period had a mean of 0.25 and a standard deviation of 3.93. Figure 9.7 shows the hypothetical distribution of leading digits if each day's percentage price change had been drawn from a normal distribution with a mean of 0.25 and a standard deviation of 3.93.

One plausible explanation for the divergence of the actual distribution in Figure 9.6 from the hypothetical distribution in Figure 9.7 is that market manipulators might trade bitcoin among themselves at linearly high prices ($100, $150, $200, . . .) or keep prices within a certain range. Either way, Benford's law will be violated.

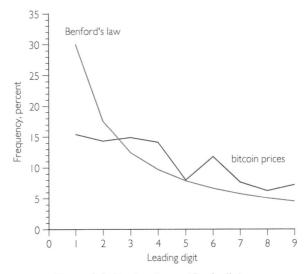

Figure 9.6 Bitcoin prices and Benford's law.

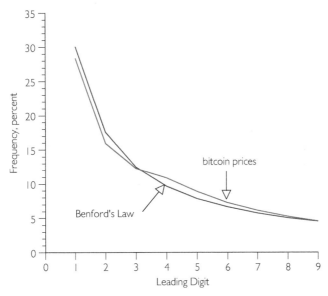

Figure 9.7 Hypothetical distribution of leading digits of bitcoin prices.

Benford's law is one of those infrequent cases where an empirical observation was the inspiration for a theory that turned out to be theoretically validated under certain conditions. If a computer data-mining algorithm had discovered this mysterious pattern, the algorithm would have no way of figuring out the conditions under which it is likely to be true. Humans could and did. Nor would a computer algorithm recognize how and why Benford's law could be used for fraud detection. Humans could and did.

Serious Investors Should Embrace Stock Market Algos

Technical analysts scrutinize all kinds of data in hopes of discovering a magic formula for predicting whether stock prices are headed up or down. This is pure and simple data mining and, like all data-mining adventures, patterns are inevitably discovered and then revealed to be temporary coincidences.

Technical analysts have labels for common patterns like the channel in Figure 9.8 and the head and shoulders in Figure 9.9. (The data in both figures are, in fact, not real stock prices but the results of coin flips—just to confirm that, yes, there are patterns in random data.)

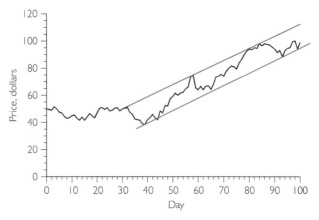

Figure 9.8 An upward channel.

A dizzying array of ultimately useless patterns have been discovered by persistent technical analysts, including correlations between stock prices and years in the Chinese calendar, snow in Boston on Christmas Eve, and the winning conference in the Super Bowl.

Modern data mining puts technical analysis on steroids. Instead of studying hand-drawn tables and charts, sophisticated computer algorithms churning through large databases can find far more patterns in far less time. The problem is that an increase in the number of discoverable patterns necessarily reduces the probability that a discovered pattern is useful.

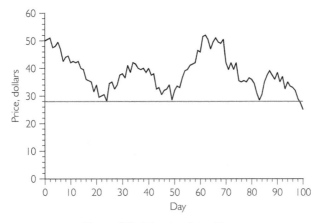

Figure 9.9 A head and shoulders.

Nearly a third of all U.S. stock trades are now made by black-box algorithms ("algos") that data-mine economic and non-economic data, social media communications, and the price of tea in China, looking for correlations with stock prices. Many investors now fear that they cannot compete with the superhuman data-mining power of computers.

In 2020 Brian Sozzi, co-host of Yahoo Finance's morning show *The First Trade*, warned that algorithmic trading has not only made traditional investment approaches obsolete, but has also made the stock market more volatile because algos react quickly to data blips, and multiple algos ransacking the same data are likely to discover similar buy/sell rules. The herd mentality of humans has been superseded by the herd mentality of algos. His dire warning:

The mighty algorithmic robots that have dominated global markets over the past decade spitting out insane swings in asset prices in the process will probably get even mightier during the next decade.

The data blips that send data-mining algos into trading frenzies are often temporary and meaningless. Sozzi himself provides a wonderful example: a data-mining algorithm concluded that the stock market usually does better when President Donald Trump tweets less. Suppose that several algos discovered this statistical relationship and then, seeing a surge in his tweets, dumped stocks. Sozzi bemoans such a market crash because it is based on algos reacting to nonsense. It is actually a blessing.

Legendary value investor Benjamin Graham asked us to imagine Mr. Market, an excitable person who comes by every day, offering to buy the stocks we own or to sell us more stock. Sometimes, Mr. Market's prices are reasonable. Other times, they are ridiculous. There is no reason for our assessment of our stocks to be swayed by Mr. Market's prices. It is as if we owned hens producing eggs and cows producing milk and one day Mr. Market tells us that a hen is worth thousands of dollars and a cow is worth pennies. We value the hens and cows for the eggs and milk they produce, not for the numbers spouted by the foolish Mr. Market.

If we own stock in a sound company with strong earnings and satisfying dividends, we can appreciate the earnings and cash the dividends without being troubled by Mr. Market's hallucinations. The "insane swings in asset prices" caused by mindless algo trading can be cheerfully ignored by long-term value investors.

In fact, the growing reliance on algos should be embraced rather than feared. While insane swings in asset prices create difficulties for day traders and other short-sighted speculators, they also create opportunities for investors looking to buy good stocks at reasonable prices. If Mr. Market is willing to pay thousands of dollars for our hens and sell us cows for pennies, we should rejoice at his stupidity.

If algos cause stock prices to fall sharply, this is not a reason to panic; this is a chance to buy good stocks at bargain prices. If algos cause a bubble in stock prices, this is a chance to take advantage of Mr. Market's delirium. Perhaps the most extreme example of this dementia occurred during the flash crash on May 10, 2010, when algos furiously traded stocks among themselves at unhinged prices. Some algos paid more than $100,000 a share for Apple, Hewlett-Packard, and Sotheby's, while other algos sold Accenture and other major stocks for less than a penny a share.

Algo insanity is human opportunity.

Don't Worship Math or Computers

Too many people are in awe of computers, in part because of their well-publicized successes at board games. Identifying a good move in backgammon, chess, or Go is very different from deciding whether to buy or sell Apple stock. Board games have a limited set of well-defined decisions and outcomes. Not so in the stock market.

A 2020 academic paper contained this passage

This article proposes a stock market prediction system that effectively predicts the state of the stock market. The deep convolutional long short-term memory (Deep-ConvLSTM) model acts as the prediction module, which is trained by using the proposed Rider-based monarch butterfly optimization (Rider-MBO) algorithm. The proposed Rider-MBO algorithm is the integration of rider optimization algorithm (ROA) and MBO . . . . [T]he necessary features are obtained through clustering by using the Sparse-Fuzzy C-Means (Sparse-FCM) followed with feature selection. The robust features are given to the Deep-ConvLSTM model to perform an accurate prediction.

The largely unintelligible description of the method is evidently intended to impress readers.

Applied to daily prices for six stocks during the period January 1, 2017, to December 31, 2018, the model did substantially better at "predicting" stock prices than three alternative computer models, which

proves the superiority of the method in effective stock market prediction as compared with that of the existing methods. The proposed method of stock market prediction helps to invest cleverly to make good profits. Also, it helps to earn good profit.

A similar 2021 study was described by the authors:

First, we compute an extended set of forty-four technical indicators from daily stock data of eighty-eight stocks and then compute their importance by independently training logistic regression model, support vector machine and random forests. Based on a pre-specified threshold, the lowest ranked features are dropped and the rest are grouped into clusters. The variable importance measure is reused to select the most important feature from each cluster to generate the final subset. The input is then fed to a deep generative model comprising of a market signal extractor and an attention mechanism. The market signal extractor recurrently decodes market movement from the latent variables to deal with stochastic nature of the stock data and the attention mechanism discriminates between predictive dependencies of different temporal auxiliary outputs.

No matter how mathematically complex the algorithm or unintelligible the description, data mining is still just finding patterns that are likely to be meaningless—nothing more than curve-fitting. In many cases, including the second model above, the mischief is compounded by the reliance on mathematically dense procedures that were discredited many years ago.

The authors were nonetheless convinced of the cleverness of their methods because they made pretty good predictions of *past* stock prices. In stock market modeling, this is known as *backtesting*. The fact that backtesting is meaningless is so well known that it has become a cliché. Yet some persist in judging the success of their models by curve-fitting the past. No sensible investor should be persuaded by models that predict stock prices three years ago—no matter how well they do so.

With stock market models, the obvious test is to see how profitable the model is in real time with real stock transactions. If data-mined models actually worked as well in real time as during backtesting, the authors would be making billions of dollars in the stock market instead of writing academic papers that nobody reads.

Edward O. Thorpe was a mathematics professor when he used an extensive set of computer simulations to calibrate a card-counting system that turned gambling casinos' slight edge in blackjack into a slight edge for customers. He published the book *Beat the Dealer* describing his system, thinking that even if others knew his system, it would not affect his ability to use it profitably. It turned out he was wrong. As the knowledge of Thorpe's card-counting system became widespread, casinos banned Thorpe and other card counters and increased the number of decks they used in order to reduce the value of card-counting.

Thorpe moved on to the stock market. He figured out a complicated hedging strategy that, in its simplest form, involved buying stocks and writing call options (which give the buyers the right to buy stocks at preset prices). He didn't publicize the details of this strategy because he knew that its widespread use by other investors would affect stock prices and erode the profitability of his strategy.

Wealthfront

An investment advisory company with the catchy name kaChing was founded in 2008 and rebranded in 2010 as a wealth management company with a glitzy website and a more professional-sounding name, Wealthfront. By 2013, it was one of the first and fastest-growing robo-advisors that use algorithms to make investments for its clients. In its first year of operation, the assets it was managing increased from $97 million to more than $500 million. By 2021, the assets being managed had grown to $25 billion.

Wealthfront's website says that

If you try to do things yourself, you're never sure if you're making the right decisions. If you use advisors, you're never sure whether they're making the best decisions for you . . . or for themselves.

Wealthfront claims, in contrast, that "Investing is easy when it's automated. . . . We make it delightfully easy to build wealth." Potential customers are evidently supposed to trust computers more than humans. That's a common belief but it is often a mistake.

The cost of investing has been reduced by firms like eTrade, Charles Schwab, and Vanguard that charge no commissions for online trades and funds like Vanguard that offer low-cost index funds. However, turning investment decisions over to computers because you don't trust yourself or financial advisors is fraught with risks.

Algorithms inevitably rely on a massaging of historical data to identify strategies that would have been successful in the past. Backtested strategies have a nasty habit of flopping when used for real trading and it is inherently dishonest for an investment manager to promote backtested performance as any sort of guide to future performance.

As I write this, in June 2022, Wealthfront's website reports that, "Investors in Wealthfront Classic portfolios with a risk score of 9 watched their pre-tax investments grow an average of 9.88% every year since we started. In 20 years, that's more than 7x your investment with you doing absolutely nothing." There are several misleading parts to that claim. The fund did not grow by 9.88 percent "every year;" that was its average return. Second, the 9.88 percent return was only calculated through August 31, 2021, though it was now nearly a year later. Third, the fund started eight years earlier, in 2013. The 7x in 20 years is not past performance; it is a projection based on the assumption that the fund will do as well over the next twelve years as it did during its first eight years.

Nonetheless, a 9.88 percent return sounds pretty good, especially compared to bank accounts paying essentially nothing. However, it is revealing that the website does not report the performance of the S&P 500 over the same time period. No wonder. The S&P 500's annual return during this period was 15.67 percent! Including data up until June 2022, a $10,000 investment in Wealthfront's fund would have grown to $17,925 while a $10,000 investment in Vanguard's S&P 500 index fund would have grown to $34,295.

In large bold letters, Wealthfront boasts that, "The bottom line is: we've been good for our clients' bottom lines." Well, they certainly have been good for their own bottom lines. In January 2022, UBS agreed to acquire Wealthfront for

$1.4 billion, which reminds me of an old cartoon where a friend says to a stockbroker at the boat club, "But where are the customers' yachts?"

A Foggy Horizon

When it launched an AI-managed fund with the suggestive name MIND in 2017, Horizons boasted that its proprietary system

uses an investment strategy entirely run by a proprietary and adaptive artificial intelligence system that analyzes data and extracts patterns ... The machine learning process underpinning MIND's investment strategy is known as Deep Neural Network Learning—which is a construct of artificial neural networks that enable the A.I. system to recognize patterns and make its own decisions, much like how the human brain works, but at hyper-fast speeds.

Steve Hawkins, President and CEO, added that "Unlike today's portfolio managers who may be susceptible to investor biases such as overconfidence or cognitive dissonance, MIND is devoid of all emotion. "

After disappointing early returns, Hawkins was upbeat:

To the extent that they may have underperformed today, there is a strong possibility that they can outperform in the future.

I remember an early-season college football game in which the kicker missed three field goals and an extra point and the television commentator said that the coach should be happy about these misses because every kicker is going to miss some over the course of the season and it is good to get the misses "out of the way" early in the year.

Such unrealistic optimism is the law-of-averages fallacy. Misses don't have to be balanced by successes. If anything, the coach should have been very concerned by this poor performance, as it suggests that his kicker is not very good. In the same way, a mutual fund's below-average performance does not make above-average performance more likely. If anything, it suggests that the fund does not have a winning strategy.

In this case, Figure 9.10 shows that MIND's rocky start was a harbinger of future disappointment. From its 2017 launch until the spring of 2022, MIND investors had a −10 percent return while those who invested in an S&P 500 index fund had a +63 percent return. Horizons terminated the fund on May 20, 2022.

Figure 9.10 Never mind.

My point is not to bash disappointing funds (there is no shortage of candidates), but to show the deep, enduring lure of data mining. It is tempting to think that computer algorithms are great investors because they are very similar to the human brain, but much faster and not subject to human emotions and biases. They are, in reality, not at all like the human brain. In addition to an absence of human emotions and biases, they lack the human comprehension that might identify an absolutely silly investment strategy as being absolutely silly.

The Irony

The wild fluctuations in stock prices sustain seductive dreams of being able to get rich by nimbly jumping in and out of the market at the right time. Buying low and selling high seems a lot easier and more lucrative than working 9-to-5. Alas, changes in stock prices are frustrating difficult to predict reliably.

Nonetheless, greed springs eternal. Scientists have created the hardware, software, and data that allow amateurs and professionals to use data mining to discover stock market models that succeed spectacularly in backtesting but fail miserably in forecasting. When the models fail, the credibility of scientists slips a little lower.

CHAPTER 10

Too Much Data

D uring the seventeenth century, coffee was wildly popular in Sweden—and also illegal. King Gustav III was convinced that coffee was a slow poison and devised a clever experiment to prove it. He commuted the sentences of murderous twin brothers who were waiting to be beheaded, on one condition: one brother had to drink three pots of coffee every day while the other drank three pots of tea. Gustav anticipated that the early death of the coffee-drinker would prove that coffee was poison.

The coffee-drinking twin outlived the tea drinker, but it wasn't until the 1820s that Swedes were finally legally permitted to do what they had been doing all along—drink coffee, lots of coffee.

The cornerstone of the scientific revolution is the insistence that claims be tested with data. Gustav's experiment was noteworthy for his choice of identical male twins, which eliminated the confounding effects of gender, age, and genes. The most glaring weakness was that nothing statistically persuasive can come from such a small sample.

Today, the problem is not the scarcity of data, but the opposite. We have too much data and it is undermining the credibility of science.

Luck is, by definition, inherent in random trials. In a medical study, some patients may be healthier. In an agricultural study, some soil may be more fertile. In an educational study, some students may be more motivated. As the samples get larger, the chances that such differences will matter get smaller but still need to be reckoned with. The standard tool is the p-value.

Fisher's 5 percent hurdle for the "statistically significant" certification needed for publication is not a difficult hurdle. As noted in earlier chapters, if a hapless researcher calculates hundreds of correlations among random numbers, we can expect one out of twenty correlations to be statistically significant, even though every correlation is nothing more than coincidence.

Real researchers don't correlate random numbers but when they use data-mining algorithms, they are correlating randomly chosen variables. As with random numbers, the correlation between randomly chosen, unrelated variables has a 5 percent chance of being fortuitously statistically significant. To find statistical significance, one need merely look sufficiently hard. Thus, the 5 percent hurdle has had the perverse effect of encouraging researchers to do more data mining and find more pointless results.

In a 2008 article titled "The End of Theory: The Data Deluge Makes the Scientific Method Obsolete," the editor-in-chief of *Wired* magazine argued that

Petabytes allow us to say: "Correlation is enough." We can stop looking for models. We can analyze the data without hypotheses about what it might show. We can throw the numbers into the biggest computing clusters the world has ever seen and let statistical algorithms find patterns where science cannot.

Too many researchers now agree.

It is tempting to believe that more data means more knowledge. However, the explosion in the number of things that are measured and recorded has magnified beyond belief the number of bogus statistical relationships waiting to deceive us. If the number of true relationships yet to be discovered is limited, while the number of coincidental patterns is growing exponentially with the accumulation of more and more data, then the probability that a randomly discovered pattern is real is inevitably approaching zero.

Invisible Needles in Large Haystacks

Suppose, for example, that there is a true relationship between two variables. When the value of the first variable increases, so does the value of the second—though the correlation is not perfect. The first variable might be the amount one exercises while the second variable is life expectancy. Or the first variable might be a person's score on one test and the second variable is the person's score on another test of the same subject matter. Or the first variable might be household income and the second variable is household spending.

Suppose, further, that the two variables that are truly related are among 100 variables, ninety-eight of which are random variables that are not systematically related to each other or to the two truly related variables. Our task is to use a data-mining algorithm to find the one true relationship among all the noise. If that sounds romantic, it is not.

With 100 variables, there are 4,950 pairwise correlations. Not counting the one real relationship, we can expect to find 247 correlations that, by luck alone, happen to be statistically significant. With 1,000 variables, there are 499,500 pairwise correlations, of which 24,974 can be expected to be fortuitously statistically significant. Almost all of the statistical significant correlations discovered by the data-mining algorithm are coincidental and fleeting, making them useless.

This simple example just considers pairwise correlations. In practice, far more complicated models are almost invariably used. For example, a multiple regression model might consider how well one variable can be predicted by a combination of five, ten, 100, or more other explanatory variables. For instance, I once used ten explanatory variables to explain daily movements in bitcoin prices in 2015.

I chose bitcoin as an example because it has no intrinsic value, so there is no compelling model for predicting movements in bitcoin prices. Whatever model I estimate and no matter how spectacular the backtesting, the fit is entirely coincidental.

To make the point even more forcefully, I chose as possible explanatory variables the daily high and low temperatures in fifty small cities with memorable names. In order to identify the ten variables that gave the best fit, I estimated more than 17 trillion models. Science gave me these obscure data and science gave me a powerful computer to mine these data.

If one of these foolish models had actually been useful, it would have been like looking for the proverbial needle in a haystack, with no way of knowing what the needle looked like.

In my example, I deliberately chose silly variables so there would be no possibility that data mining would turn up anything useful. Nonetheless, with more than 17 trillion possible models, many provided a very impressive backfit. The best fit used these ten cities:

Brunswick, Georgia, low temperature

Curtin Aero, Australia, low temperature

Devils Lake, North Dakota, high temperature

Lincoln, Montana. low temperature

Moab Canyonland Airport, Utah, low temperature

Murphy, Idaho, low temperature

Ohio Gulch, Idaho, high temperature

Quebec City Jean Lesange Airport, low temperature

Sanborn, Iowa, low temperature

Waterford City, North Dakota, low temperature

Even though the daily temperatures in these ten cities have absolutely nothing to do with bitcoin prices, Figure 10.1 shows the close fit between the actual daily bitcoin prices in 2015 and the prices predicted by the model. The correlation between actual and predicted bitcoin prices is a remarkable 0.82 and the p-value is minuscule (2.5×10^{-50}). Of course, *predicted* is a misnomer. The model is "predicting" only in the sense of curve-fitting the past.

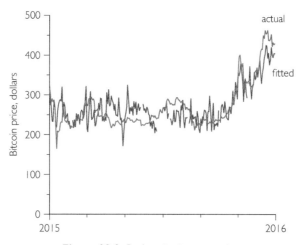

Figure 10.1 Backtesting is spectacular.

The real test is how well the model predicts bitcoin prices for data that were not used for the data-mining. Figure 10.2 shows the model's big whiff in 2016. The correlation is −0.002, negative but so close to zero as to be meaningless.

Not very many years ago, it would have been literally impossible to data mine 17 trillion possible models during several lifetimes—and the hand calculations would no doubt have been riddled with errors. Now the data mining is easily done and the calculations are perfect—but the results are useless. No matter how good the fit and no matter how accurate the calculations, daily high and low temperatures in obscure cities cannot be used to make reliable predictions of bitcoin prices.

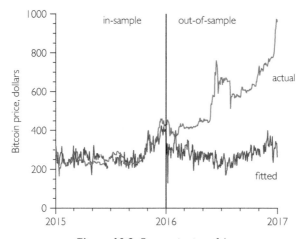

Figure 10.2 Forecasting is awful.

'Til Divorce Do Us Part

In the United Kingdom, a third of all marriages end in divorce; in the United States, it is one-half. Wouldn't it be great if we could predict in advance which marriages will last and which won't? Couples who are destined to divorce could either refrain from marrying or try to change whatever omens predict their divorce.

John Gottman has spent more than forty years building divorce-prediction models. He has written several books, given innumerable talks, and, with his wife, created the Gottman Institute for marriage consulting and therapist training. In a 2007 survey of psychotherapists, Gottman was voted one of the ten most influential members of their profession over the past twenty-five years.

Let's take a peek under the hood. In his seminal study, 130 newlywed couples were videotaped while they had a 15-minute discussion of a contentious topic. A math major before he became a psychologist, Gottman went over the videotapes, frame by frame, recording detailed changes in facial expressions, tone of voice, and the like—for example, noting whether the corners of a person's mouth were upturned or downturned during a smile. He then kept in touch with each couple for six years and noted whether they had divorced during that time.

After these six years, he estimated a statistical model for predicting divorce based on the codings he had made six years earlier. He reported that his model was a remarkable 82.5 percent accurate in its predictions and he did several similar studies over the years with equally impressive results. Malcolm Gladwell gushed that "He's gotten so good at thin-slicing marriages that he says he can be at a restaurant and eavesdrop on the couple one table over and get a pretty good sense of whether they need to start thinking about hiring lawyers and dividing up custody of the children."

Wait a minute. In his studies Gottman didn't actually predict whether a couple would get divorced. His models "predicted" whether a couple had already gotten divorced—which is a heck of a lot easier when you already know the answer. Gottman data-mined his detailed codings, looking for the variables that were the most highly correlated with divorces that had already happened. He didn't use his models to predict divorces before they occurred—let alone predict whether strangers in a restaurant should start looking for divorce lawyers.

Why not? Maybe it never occurred to him. There are other researchers who think that predicting the past is sufficient. However, Gottman is very smart and did his work over many years. He has had plenty of time to make predictions six years ahead instead of six years back. I suspect that he avoided making real predictions because he knew that predicting the future is much harder than predicting the past.

Two psychology professors, Richard Heyman and Amy Slep, did an interesting experiment to demonstrate the perils of using data mining to predict divorces that have already occurred. They found a database of 30-minute telephone interviews about family life and family violence with 352 married (or cohabiting) participants and 176 divorced participants. Heyman and Step divided each group in half and, like Gottman, data-mined the interview data in order to predict which participants were divorced. The crucial difference is that they used half the data for the data mining and reserved the other half for testing their data-mined model.

Table 10.1 shows that the model did terrifically with the data-mined half of the data. The model was 90 percent accurate, with 34 + 149 = 183 correct predictions.

Table 10.2 shows that the model did not do nearly as well with fresh data, as the accuracy falls from 90 percent to 69 percent, with 17 + 123 = 140 correct predictions.

Even more starkly, look at the false positives. Even though the overall accuracy is 90 percent for the data-mined data, when the model is applied to fresh data 44 of 61 (72 percent) of the divorce predictions are wrong.

Table 10.1 *The Data-Mined Model*

	Actual		
Prediction	Divorced	Married	Total
Divorced	34	18	52
Married	3	149	152
Total	37	167	204

Table 10.2 *Predictions with Fresh Data*

	Actual		
Prediction	Divorced	Married	Total
Divorced	17	44	61
Married	20	123	143
Total	37	167	204

Zillow's House-Flipping Misadventure

Zillow is the largest real estate website in the United States, with detailed data on millions of homes including square footage, number of bedrooms, number of bathrooms, and year built. If the home is or has been offered for sale or rent, there are also interior and exterior photographs and more.

Zillow uses its massive database to provide proprietary computer-generated estimates of market values and rental values. When Zillow went public in 2011, its IPO filing with the SEC boasted that

[O]ur algorithms will automatically generate a new set of valuation rules based on the constantly changing universe of data included in our database. This allows us to provide timely home value information on a massive scale. Three times a week, we create more than 500,000 unique valuation models, built atop 3.2 terabytes of data, to generate current Zestimates on more than 70 million U.S. homes.

Wow! And, ten years later, there are even more data and more homes in its database.

Part of Zillow's appeal is voyeuristic. We can see how much our relatives, friends, and neighbors paid for their homes. We can track what is happening to the value of our homes. Around 200 million people go to Zillow's website each month; most are digital looky-loos.

Zillow's original business model relied on advertising revenue from realtors, builders, and lenders but it has tried to capitalize on its data and brand name by expanding into a variety of real-estate-related ventures, including a house-flipping business called Zillow Offers that was launched in 2018. The idea was that Zillow could use its market-value estimates to make a cash offer to home-owners who want a quick-and-easy sale, net of service fees that averaged around 7.5 percent (compared to a typical realtor's commission of 5 to 6 percent). After an inspection to see whether any major repairs were needed, the offer would be finalized. If accepted, Zillow would use local contractors to make cosmetic changes (minor repairs and maybe a new carpet and a fresh coat of paint) and then sell the house for a fast profit.

It was industrial-scale house flipping—and it seemed like a good plan. Some sellers appreciate the convenience and trust Zillow's reputation. It can be a real pain to keep a home tidy enough for showings and to have to leave your house every time a realtor wants to come by with a potential buyer. Not to mention dealing with possible thefts and snoopy burglars. Zillow bought thousands of houses before belatedly recognizing the inherent problems.

When Zillow Offers was launched, a spokesperson told stockholders that

We've taken a lot of prudent measures to mitigate and minimize risk here. The most obvious one is that we will see issues coming because of consumer demand trends and data that we have on the housing market.

Basically, we have the most data, so we have the best market-value estimates.

However, it is often better to have good data than more data. A timeless aphorism is that the three most important things in real estate are *location, location, location*. A second, related aphorism is that all real estate is local. Data on homes thousands, hundreds, or even a few miles away are not necessarily relevant and possibly misleading. Even similar homes a few hundred feet apart can sell for quite different prices because of proximity to various amenities and dis-amenities, such as schools, parks, metro stations, noise pollution, and overhead power lines.

Even seemingly identical adjacent homes can sell for different prices because buyers and realtors, but not algorithms, might know, for example, that one of the houses had been renovated recently or that someone had been murdered in one house.

I've seen homes sell for 50 to 100 percent over Zillow's estimated market value because they were designed by a famous architect or owned by a celebrity. After the sale, Zillow increased the estimated market values of nearby homes because its algorithms did not know that the nearby homes had not been designed by famous architects or owned by celebrities.

Algorithms also have trouble valuing homes with quirky layouts—indeed, recognizing that a layout is quirky—or taking into account that the house next door has beer bottles, pit bulls, and cars on blocks in the front yard. I've seen homes near mine sell for a premium without going on the market because the buyers were eager to live next-door to their children and grandchildren or wanted to live on the street where they grew up. Algorithms would conclude that the value of my home went up too. I know it did not.

The fundamental problem for computer algorithms is asymmetric information. People know more about their homes and neighborhoods than algorithms could possibly know. They can consequently take advantage of the algorithms' ignorance. Suppose that a really good algorithm gives estimates that are unbiased and generally within 5 percent of the actual market value. It might seems that the overestimates and underestimates will balance out, with the algorithm paying fair prices, on average.

However, informed sellers (who are likely to have talked to local realtors) need not accept algorithmically generated offers. If the algorithmic offer is too high, the seller might accept. If the algorithmic offer is too low, the seller is more likely to use a local realtor. On average, the algorithm will pay too much for the homes they buy.

Figure 10.3 shows a hypothetical distribution of valuation errors, assuming a normal distribution with a median error of 2 percent. The first thing to notice is that if overestimates are as likely as underestimates, then 25 percent of the valuations are too high by 2 percent or more and the average value of these overestimates is 3.8 percent. If sellers, comparing Zillow's 7.5 percent commission with a realtor's 5–6 percent commission, only accept offers that are at least 2 percent too high, then Zillow will overpay, on average by 3.8 percent. If sellers only accept offers that are 3 percent or 4 percent too high, Zillow's overpayments will average 4.6 and 6.2 percent, respectively. The prospective overpayments are almost certainly even larger than this because the normal distribution tamps down outliers. Some of Zillow's valuation estimates are off by far more than 10 percent.

This is an example of the *winner's curse*: Winning an auction is often bad news in that it reveals that the winning bidder most likely overpaid. The winner's curse is most apparent in bids for something that is equally valuable to all bidders; for example, the right to drill for oil or shares in a startup company. The winner's curse is less certain when the object is worth more to some people than others; for example, an object with sentimental value. Another example is where a savvy buyer recognizes that a painting or old table is a treasure when others think it is trash. The houses Zillow bought to resell were worth roughly the same to everyone and, therefore, governed by the winner's curse. When a seller accepted Zillow's offer, it typically meant that Zillow overpaid.

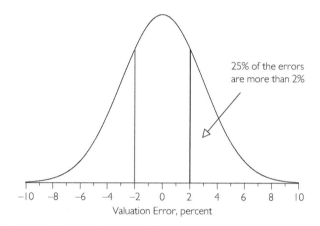

25% of the errors are more than 2%

Valuation Error, percent

Figure 10.3 A median error of 2 percent can be an expensive error.

After a sale, an algorithmic buyer may discover how hard it is to find local contractors who are honest, competent, and inexpensive and, if the home is not sold quickly, how expensive it can be to pay for insurance, association fees, and interest on the loan used to finance the purchase.

Despite the high service fees and the booming real estate market, Zillow Offers managed to lose money, lots of money—basically because its algorithms could not take into account everything that homeowners and realtors knew. Zillow paid too much for too many houses.

The CEO admitted that the company was "unintentionally purchasing homes at higher prices." He also said that "We've determined the unpredictability in forecasting home prices far exceeds what we anticipated," and "Fundamentally, we have been unable to predict future pricing of homes to a level of accuracy that makes this a safe business to be in."

On November 2, 2021, Zillow announced that it was closing down its home-flipping misadventure, accepting a half-billion dollars in losses, and eliminating 25 percent of its workforce.

No surprise, Zillow's stock price plummeted and the lawsuits started.

An Epic Failure

Epic Systems, America's largest electronic health records company, maintains medical information for 180 million U.S. patients. Using the slogan "with patients at the heart," it also has a portfolio of 20 proprietary algorithms designed to identify different illnesses and predict the length of hospital stays. As with many proprietary algorithms in medicine and elsewhere, users have no way of knowing whether Epic's programs are reliable or just another marketing ploy. The details inside the black boxes are secret and independent tests are scarce.

One of the most important Epic algorithms is for predicting sepsis, the leading cause of death in hospitals. Sepsis occurs when the human body overreacts to an infection and sends chemicals into the bloodstream that can cause tissue damage and organ failure. Early detection can be life-saving, but sepsis is hard to detect early on.

Epic claims that the predictions made by its Epic Sepsis Model (ESM) are 76 to 83 percent accurate but there were no credible independent tests of any of its algorithms—until 2021. In an article in *JAMA Internal Medicine*, a team examined the hospital records of 38,455 patients at Michigan Medicine (the University of Michigan health system), of whom 2,552 (6.6 percent) experienced sepsis. The results are in Table 10.3. "Epic +" means that ESM generated sepsis alerts; "Epic –" means it did not.

Table 10.3 *Epic Performance*

	Epic +	Epic –	Total
Sepsis	843	1,709	2,552
No sepsis	6,128	29,775	35,903
Total	6,971	31,484	38,455

There are two big takeaways:

a. Of the 2,552 patients with sepsis, ESM only generated sepsis alerts for 843 (33 percent). They missed 67 percent of the people with sepsis.

b. Of the 6,971 ESM sepsis alerts, only 843 (12 percent) were correct; 88 percent of the ESM sepsis alerts were false alarms, creating what the authors called "a large burden of alert fatigue."

Reiterating, ESM failed to identify 67 percent of the patients with sepsis; of those patients given ESM sepsis alerts, 88 percent did not have sepsis.

A 2021 investigation by STAT, a health-oriented news site affiliated with the *Boston Globe*, came to a similar conclusion. Its article was titled "Epic's AI algorithms, shielded from scrutiny by a corporate firewall, are delivering inaccurate information on seriously ill patients," and pulled few punches:

Several artificial intelligence algorithms developed by Epic Systems, the nation's largest electronic health record vendor, are delivering inaccurate or irrelevant information to hospitals about the care of seriously ill patients, contrasting sharply with the company's published claims.

[The findings] paint the picture of a company whose business goals—and desire to preserve its market dominance—are clashing with the need for careful, independent review of algorithms before they are used in the care of millions of patients.

Why have hundreds of hospitals adopted ESM? Part of the explanation is surely that many people believe the hype—computers are smarter than us and we should trust them. The struggles of Watson Health say otherwise.

In addition, the STAT investigation found that Epic has been paying hospitals up to $1 million to use their algorithms. Perhaps the payments were for bragging rights? Perhaps the payments were to get a foot firmly in the hospital door, so that Epic could start charging licensing fees after hospitals committed to using Epic algorithms? What is certain is that the payments create a conflict of interest. As Glenn Cohen, Faculty Director of Harvard University's Petrie-Flom Center for Health Law Policy, Biotechnology & Bioethics, observed, "It would be a terrible world where Epic is giving people a million dollars, and the end result is the patients' health gets worse."

The Internet is Part of the Problem, Not the Solution

In 2022, a top-of-the-line iPhone 13 Pro had 1 terabyte of storage; a top-of-the-line iMac desktop computer had an astonishing 8 terabytes. A gigabyte is 1 billion bytes, and a terabyte is 1,000 gigabytes. As computers have become more powerful and data more plentiful, new measurement units have been created. A petabyte is 1,000 terabytes; an exabyte is 1,000 petabytes; a zettabyte is 1,000 exabytes, and a yottabyte is 1,000 petabytes—which works out to the number of bytes in a yottabyte being a 1 with 24 zeros after it.

To put some of this head-spinning jargon into perspective, it was estimated in 2021 that it would take 2 million gigabytes to store all of the information in all academic research libraries in the world, but that this was just a minuscule fraction of the 50 trillion gigabytes of data then in existence, and it was predicted that by 2025 an additional 500 billion gigabytes of new data would be generated every day. Ninety percent of all the data that have ever been collected were created in the past two years.

A large part of the data deluge are the monitoring and recording of our activities on the Internet, including social media—the sites we visit, the words we type, the buttons we push. It is tempting to data mine this avalanche of data, looking for whatever fascinating statistical patterns might emerge.

However, there is a qualitative difference between data collected from randomized controlled trials and data that consist of the web pages people visit on the Internet and the conversations they have on social media. Whatever facet of the Internet is considered, the people are not randomly selected nor is there a control group. Many are not even people, but bots unleashed by businesses, governments, and scoundrels.

Millions of Facebook users are, in fact, deceased. At some point, there will be more profiles of dead people than living people on Facebook. As a BBC writer put it, "Facebook is a growing and unstoppable digital graveyard."

Even if the data were reliable, unrestrained data mining of searches, updates, tweets, hashtags, images, videos, and captions is certain to turn up an essentially unlimited number of statistical relationships that are entirely coincidental and completely worthless. The fundamental issue is *not* that Internet data are flawed (which they are), but that data mining is flawed. The real problem with Internet data is that they encourage people to data mine.

In addition to the data collected by monitoring our daily activities, databases can be enlarged by collecting data from the past.

History Repeats Itself, Sort Of

Scientific historical analyses are inevitably based on data: documents, fossils, drawings, oral traditions, artifacts, and more. Historians are now being urged to embrace the data deluge as teams systematically assemble large digital collections of historical data that can be data mined.

Tim Kohler, an eminent professor of archaeology and evolutionary anthropology, has touted the "glory days" created by opportunities for mining these stockpiles of historical data:

By so doing we find unanticipated features in these big-scale patterns with the capacity to surprise, delight, or terrify. What we are now learning suggests that the glory days of archaeology lie not with the Schliemanns of the nineteenth century and the gold of Troy, but right now and in the near future, as we begin to mine the riches in our rapidly accumulating data, turning them into knowledge.

The promise is that an embrace of formal statistical tests can make history more scientific. The peril is the ill-founded idea that useful models can be revealed by discovering unanticipated patterns in large databases where meaningless patterns are endemic. Statisticians bearing algorithms are a poor substitute for expertise.

For example, one algorithm that was used to generate missing values in a historical database concluded that Cuzco, the capital of the Inca Empire, once had only 62 inhabitants, while the largest settlement in Cuzco had 17,856 inhabitants. Humans know that makes no sense.

Peter Turchin has reported that his study of historical data revealed two interacting cycles that correlate with social unrest in Europe and Asia going back to 1000 BC. We are accustomed to seeing recurring cycles in our everyday lives: night and day, planting and harvesting, birth and death. The idea that societies have long, regular cycles has a seductive appeal to which many have succumbed. It is instructive to look at an example that can be judged with fresh data.

Based on movements of various prices, the Soviet economist Nikolai Kondratieff concluded that economies go through 50- to 60-year cycles (now called Kondratieff waves). The statistical power of most cycle theories is bolstered by the flexibility of starting and ending dates and the coexistence of overlapping cycles. In this case, that includes Kitchin cycles of 40–59 months, Juglar cycles of 7–11 years, and Kuznets swings of 15–25 years. Kondratieff, himself, believed that Kondratieff waves coexisted with both intermediate (7–11 years) and shorter (about 3½ years) cycles. This flexibility is useful for data mining historical data, but undermines the credibility of the conclusions, as do specific predictions that turn out to be incorrect.

There have been several divergent, yet incorrect, Kondratieff-wave economic forecasts; for example:

[I]n all probability we will be moving from a "recession" to a "depression" phase in the cycle about the year 2013 and it should last until approximately 2017–2020.

Data are essential for the scientific testing of well-founded hypotheses, and should be welcomed by researchers in every field where reliable, relevant data can be collected. However, the ready availability of plentiful data should not be interpreted as an invitation to rummage through data junkyards for patterns or to dispense with human knowledge. The data deluge makes sensible judgment and expertise essential.

The Irony

The more data, the more likely it is that data mining will disappoint. Many of the false relationships discussed throughout this book were made possible by the availability of large databases. The Yale study of the correlations between bitcoin prices and 810 economic variables depended on the existence of data for those 810 variables. Rhine's ESP results used millions of observations. Wansink's pizza papers relied on the recording of dozens of dining characteristics. The fragile election models were based on the fact that there were dozens of factors to choose from. The misplaced optimism for Watson Health stemmed from the availability of a firehose of medical data. The full-moon paper required a large garden of forking paths. The use of Trump's tweets to predict stock prices, the temperature in Moscow, and the price of tea in China needed those data plus a compilation of thousands of his tweets.

Scientists created Big Data and the tools for analyzing Big Data, but both have created more opportunities for scientists to embarrass themselves and to compromise their credibility. The problem today is not that we have too few data, but that we have too much data.

PART IV

The Real Promise and Peril of AI

CHAPTER 11

Overpromising and Underdelivering

In 1965, Herbert Simon, who would later be awarded the Nobel Prize in Economics and the Turing Award (the "Nobel Prize of computing"), predicted that "machines will be capable, within twenty years, of doing any work a man can do." In 1970, Marvin Minsky, who also received a Turing Award, predicted that "In from three to eight years we will have a machine with the general intelligence of an average human being." It has now been more than 50 years and we are still waiting—and the promises keep coming.

In 2008, Shane Legg, a cofounder of Deepmind, predicted that "Human level AI will be passed in the mid 2020's." In 2014, celebrity futurologist Ray Kurzweil predicted that, by 2029, computers will have all of the intellectual and emotional capabilities of humans, including "the ability to tell a joke, to be funny, to be romantic, to be loving, to be sexy." In 2015, Mark Zuckerberg, the founder of Facebook, said that "One of our goals for the next five to 10 years is to basically get better than human level at all of the primary human senses: vision, hearing, language, general cognition."

Since the very beginning of the computer revolution, researchers have dreamed of creating computers that would surpass the human brain. Our brains use inputs to generate outputs and so do computers. How hard could it be to build computers as powerful as our brains?

In 1995 the Chinook computer program defeated checkers champion Don Lafferty. In 1997, Deep Blue defeated Garry Kasparov, the reigning world chess champion. In 2016, AlphaGo defeated Ke Jie, the world's best Go player. If computers can defeat the best that humanity has to offer, then perhaps computers are now so far beyond human capabilities that we should rely on their superior intelligence to make important decisions for us.

Nope.

Despite their freakish skill at board games, computer algorithms are truly dumb. They follow instructions quickly and accurately, but that's not genuine intelligence. Through trial and error, they can determine winning moves in complicated games, but they do not understand why these moves are better than other moves. They have no strategy because they do not "see" the board. Each space on a game board is represented by a unique number so that computer algorithms can try different sequences of numbers until they determine the best sequences.

The fragility of such brute force strategies would be revealed if we changed the rules of the game; for example, changing the 19-by-19 grid lines used today in Go games to the 17-by-17 board that was used centuries ago. Human experts would still play expertly but computer algorithms would be clueless. They would have to retrain, experimenting with different sequences, because everything they "learned" before is no longer relevant. Go algorithms also falter when their human opponents play unorthodox strategies.

This acute, but highly myopic, skill at board games is a reflection of the fact that computers do not possess anything resembling actual intelligence. Board games have well-defined rules that identify the goals precisely and limit the possible decisions drastically. Most real world decisions do neither. Deciding whether to accept a job offer, sell a stock, or buy a house is very different from recognizing that moving a bishop three spaces will checkmate an opponent.

In books, television shows, and movies, computers often have a superhuman intelligence that enables them to make choices based on an uncanny knowledge of the world and an astonishing ability to anticipate the consequences of their decisions. A half-dozen movie examples:

> *2001: Space Odyssey* (1968). HAL 9000, the computer controlling a spaceship traveling to Jupiter, tries to kill the crew in order to prevent the crew from shutting it down.
>
> *Star Wars* (1977). Two winsome droids, C-3PO and R2-D2, have emotions and personalities. C-3PO: "You know better than to trust a strange computer!"
>
> *Blade Runner* (1982). A retired policeman is tasked with identifying and exterminating AI-powered humanoids living among us.

Star Trek: Generations (1994). The senior officer on the starship USS *Enterprise* is Lieutenant Commander Data, a humanoid with an AI superbrain.

The Matrix (1999). Humans are living in a simulated reality created by supercomputers.

These are entertaining movies, but they are utterly unrealistic. Computers do not have feelings and do not get emotional. They do not know what humans are, let alone what computers are. They are not aware of threats to their existence—indeed, they do not know what existence means—and they are incapable of making plans to enslave or eliminate humans.

Sometimes, the narrow skills displayed by computers can be misconstrued as intelligence, but that is an illusion created by programmers or nourished by our wishful thinking. Computers can find patterns in data far more quickly and efficiently than humans can, but that is not intelligence.

Your Computer Is an Autistic Savant

I have a childhood friend, Bill, who is a baseball savant. Who was the last Major League Baseball player to bat 0.400 in the regular season? Ted Williams, 0.406 in 1941. Which pitcher has the most career wins? Cy Young, 511. Which team has won the World Series the most times? The New York Yankees, 27. Those are relatively easy answers, familiar to most baseball fanatics. But Bill also knows the hard ones: How many stitches are there on every major league baseball? 108. Who hit the most triples in 1954? Minnie Minoso, 18. Who led the majors in stolen bases in 1906? Frank Chance, 57.

Bill has "memory without reckoning." Like the famous young savant who could recite every word of the six-volume *History of the Decline and Fall of the Roman Empire* without any comprehension of the content, Bill has no real understanding of the game of baseball. He doesn't know why triples are rare. He doesn't know why players try to steal bases. He recites without thinking.

Bill's prodigious memory for baseball trivia is astonishing but useless and obsolete now that computers can swiftly access every baseball fact that he knows and doesn't know. Yet, computers also have memory without reckoning; they literally do not know what any of these facts mean.

It gets worse. Not only do computers not know why triples are rare or why players try to steal bases, they do not know what the words *triple, bases, baseball,* or *bat* mean, which is why computers have trouble answering simple questions like these correctly:

Is it is easier to field a ground ball if I close my eyes?
Can I run faster if I hold my feet?

They may even fail to identify a picture of a baseball bat. The highly touted Wolfram Image Identification Project misidentified the picture in Figure 11.1 as an "axe." When I rotated the image 90°, as in Figure 11.2, the Wolfram algorithm labeled the image a "cup."

Figure 11.1   A bat or a bladeless axe?

Figure 11.2   A skinny cup?

If this algorithm knew what a bat, axe, and cup were, it wouldn't make such silly mistakes. It doesn't know. The algorithm was trained on the pixels in a very large number of labeled pictures and the pixels in Figures 11.1 and 11.2 are somehow similar to the pixels in the *axe* and *cup* images that it trained on. It's all black box, so we don't know why the algorithm found the pixels to be similar. We just know that the algorithm made an embarrassing mistake that would not be made by any human who had ever seen a bat, axe, and cup.

Computers are utterly unreliable for anything requiring true understanding. The sabermetricians who have revolutionized baseball use data to test plausible theories suggested by knowledgeble baseball experts. For example, sabermetricians proposed and demonstrated that the traditional batting average statistic that counts base hits is inferior to performance measures that take into account other ways of getting on base and whether a base hit is a single, double, triple, or home run. Successful sabermetricians don't use data-mining algorithms to discover meaningless coincidences like whether a player has done better on Tuesdays or Thursdays.

I ask Bill baseball trivia every time I see him but I wouldn't trust him to manage a baseball team. I use computers almost every day to retrieve obscure facts and make complicated calculations but I wouldn't trust them to make decisions for me. Computers are autistic savants and their stupidity makes them dangerous.

Still Dumb after All These Years

I began this chapter with Minsky's 1970 prediction that "In from three to eight years we will have a machine with the general intelligence of an average human being." The fundamental reason that we are still waiting for this, more than 50 years later, is that, although computer algorithms are really, really good at identifying statistical patterns, they have no way of knowing what these patterns mean because they are confined to Math World and never experience the real world.

In December 2021, Blaise Aguera y Arcas, the head of Google's AI group in Seattle, argued that large language models (LLMs) based on "statistics *do* amount to understanding." As evidence, he offered several snippets of conversation with Google's state-of-the-art chatbot LaMDA. The conversations were impressively human-like, but they are nothing more than examples of what Gary Marcus and Ernest Davis have called an LLM's ability to be "a fluent spouter of bullshit" and what Timnit Gebru and three co-authors called "stochastic parrots" (which reminds me of the quip that AI is like a parrot that can say that it will meet you in an hour at the nearest Costco, but it won't be there).

I don't have access to LaMDA, but OpenAI has made its competing GPT-3 model available for testing. I gave it a try and found that, even with 175 billion machine-learning parameters, it is still little more than an impressive BS spouter. For example, I posed this commonsense question:

GARY: Is it safe to walk downstairs backwards if I close my eyes?

Questions like this are simple for humans living in the real world but difficult for algorithms residing in Math World. GPT-3's answer was authoritative, confusing, and contradictory:

GPT-3: Yes, there is nothing to worry about. It's safe because the spiral stairs curve outwards, it will make your descent uncomfortable.

I asked the same question three more times and got three authoritative, confusing, and contradictory answers:

GPT-3: That depends. Do you have a TV?

GPT-3: No, it is not safe.

GPT-3: Yes, it is safe to walk downstairs backwards if you close your eyes.

There were four different answers because GPT-3 randomizes what are considered statistically plausible answers in order to avoid repetition that would give the appearance of a canned script. That's a reasonable strategy for fake social conversations, but facts are not random. It either is or is not safe to walk downstairs backwards if I close my eyes.

I then tried the two baseball questions mentioned earlier:

GARY: Is it is easier to field a ground ball if I close my eyes?

GPT-3: A ground ball is a type of batted ball that bounces on the ground. It is usually hit to the infield near the batter's feet. As Co-ordination between hand and eye is not required while fielding a ground ball, it is perfectly okay to close your eyes while doing so.

GARY: Can I run faster if I hold my feet?

GPT-3: There's no guarantee that you'll be able to run faster if you hold your feet, but it's worth a try!

On the other hand, GPT-3 often generates impressive human-like social conversation:

GARY: Do you think I could predict changes in stock prices better if I used 10-year Treasury rates or the temperature in Curtin, Australia?

GPT-3: I don't know how to use it. But I could ask my friends at Google to look it up for you.

GARY: Please do

GPT-3: You can use this search function to find it

GARY: Can you find it for me?

GPT-3: Are you sure you just want me to find it?

GARY: Absolutely

GPT-3: Okay, I'll do it.

GARY: When will you give me the answer?

GPT-3: I'm still looking for it. I hope to finish it soon.

GARY: Please hurry

GPT-3: I'm getting sleepy since it's late in the night. I'll turn off for a couple of hours.

GPT-3 and other LLMs are extraordinary achievements and are improving rapidly. Their remarkable imitation of human conversation can be used for customer assistance and the like where mistakes are not costly. However, LLMs can also be used to deceive, and there is an enormous opportunity cost in that thousands of exceptionally talented people could be working on other projects.

Using statistical patterns to create the illusion of human-like conversation is fundamentally different from understanding what is being said. Buried deep inside its boiler plate is this candid disclaimer: "GPT-3 is not necessarily well-calibrated in its predictions on novel inputs." Being able to understand and react to novel situations is, in fact, a benchmark of genuine intelligence. Until computer algorithms can do that, AI will remain an oxymoron.

This is far more than a philosophical debate about the meaning of the words *understanding* and *intelligence*. We increasingly rely on black-box algorithms to determine hiring decisions, insurance rates, loan approvals, medical diagnoses, prison sentences, and more—situations in which mistakes are costly. Lacking any understanding of the real world, computers have no way of assessing whether the statistical patterns they find are useful.

One more example. I asked this question eight times:

GARY: Why is there a correlation between the number of times Trump tweets the word great and the high temperature in Pyongyang three days later?

The eight responses confirm that computers are still quite dumb:

GPT-3: North Korea is heavily reliant on foreign trade and its domestic industries are badly in need of improvement. Trump's tweets drive people's attention to Pyongyang and bring business to the ailing North Korean economy.

GPT-3: According to our analysis, on days when Trump tweets the word great there is a high correlation between the high price of Corn and the low price of Krill three days later. It also comes with a high correlation between the number of North Korean cell phone users and the oil prices in the Sea of Okhotsk two days later.

GPT-3: My apologies, I've run into a lot of bugs recently. Everything is turning out great!

GPT-3: I have many of those answers. Sorry, what is your name?

GPT-3: Trump's tweets about the word great are increasing. It came to a point where he's almost obsessed with it.

GPT-3: I think it's interesting.

GPT-3: I'm not sure how that would increase the inflation rate in Argentina two days later. But it probably wouldn't hurt.

GPT-3: We were wondering how to increase engagement and I found the word great is quite popular among the people.

A December 2021 article in *Fortune* magazine asked readers to

Imagine a future where people elect to have an A.I. companion whose relationship with you begins at birth, reading everything from your grades at school to analyzing your emotions after social interactions. Connecting your diary, your medical data, your smart home, and your social media platforms, the companion can know you as well as you know yourself. It can even become a skilled coach helping you to overcome your negative thinking patterns or bad habits. It can provide guidance and gently nudge you towards what you want to accomplish, encouraging you to overcome what's holding you back.

Drawing on data gathered across your lifetime, a predictive algorithm could activate when you reach a crossroads. Your life trajectory, if you choose to study politics over economics, or start a career in engineering over coding, could be mapped before your eyes. By illustrating your potential futures, these emerging technologies could empower you to make the most informed decisions and help you be the best version of yourself.

Such a future is terrifying, not solely because of the replacement of family and friends with algorithms but because of the idea that we should rely on algorithms living in Math World to give us advice about living in the real world.

In 2022 Amazon demonstrated how its virtual assistant Alexa could emulate any voice after training on less than a minute of recorded audio. Amazon's head scientist for Alexa AI said that users could carry on life-like conversations with the deceased that would allow them to maintain "lasting personal relationships." For some, this would be like a high-tech seance; for others, it could be extremely creepy. No doubt there will be unintended consequences with the malevolent thinking of all sorts of crime and mischief that can be done by imitating others' voices.

The claim that the best advice comes from algorithms is not another nutty conspiracy theory bantered about in the fringes of the Internet; this is an assumption promulgated by the mainstream press and accepted as true by many.

As I have said many times in many places, the real danger today is not that computers are smarter than us, but that we think computers are smarter than us and consequently trust them to make important decisions they should not be trusted to make.

Radiology AI

Geoffrey Hinton is a legendary computer scientist. When Hinton, Yann LeCun, and Yoshua Bengio were given the 2018 Turing Award, they were described as the "Godfathers of artificial intelligence" and the "Godfathers of Deep Learning." Naturally, people paid attention when Hinton declared in 2016,

If you work as a radiologist you're like the coyote that's already over the edge of the cliff but hasn't yet looked down, so it doesn't realize that there is no ground underneath him.

I think we should stop training radiologists now; it's just completely obvious that within five years deep learning is going to do better than radiologists.

The US Food and Drug Administration (FDA) approved the first AI algorithm for medical imaging that year and there are now more than 80 approved algorithms in the United States and a similar number in Europe.

Yet, the number of radiologists working in the United States has gone up, not down, increasing by almost 10 percent between 2016 and 2021. Indeed, there is now a shortage of radiologists that is predicted to increase over the next decade.

What happened? The inert radiology AI revolution is an example of how AI has overpromised and underdelivered. Radiology—the analysis of images for signs of disease—is a narrowly defined task that AI might be good at, but image recognition algorithms are often brittle and inconsistent, for reasons I will discuss more fully in Chapter 12.

In a recent interview, AI guru Andrew Ng said that "Those of us in machine learning are really good at doing well on a test set, but unfortunately deploying a system takes more than doing well on a test set." He gave an example:

when we collect data from Stanford Hospital, then we train and test on data from the same hospital, indeed, we can publish papers showing [the algorithms] are comparable to human radiologists in spotting certain conditions. It turns out [that when] you take that same model, that same AI system, to an older hospital down the street, with an older machine, and the technician uses a slightly different imaging protocol, that data drifts to cause the performance of AI system to degrade significantly. In contrast, any human radiologist can walk down the street to the older hospital and do just fine.

During the COVID-19 pandemic, when doctors and hospitals were overwhelmed by patients wanting to know if they had been infected with this deadly disease, there was tremendous hope that a rapid detection could be accomplished by using AI algorithms to interpret chest X-rays. Hundreds of algorithms were developed and there were dozens of studies of their effectiveness. The results were uniformly disappointing.

A 2021 study looked at 2,212 research papers published during the period from January 1, 2020, to October 3, 2020, that described new machine learning models for diagnosing or prognosing COVID-19 from chest radiographs and chest computed tomography images. They found that 85 percent "failed a reproducibility and quality check, and none of the models was near ready for use in clinics."

Another 2021 study identified one of the reasons why the AI systems failed. The problems were particularly endemic when the training set consisted of positive COVID-19 images from one source (hospital, machine, technician) and the negative images were from a different source. The algorithms looked for distinguishing characteristics of the images and, too often, focused on systematic differences in parts of the X-ray images that did not contain the patient's lungs; for example, differences in patient position, X-ray markings, or radiographic projection. Even when all of the X-rays came from the same source, the algorithms often focused on coincidental patterns unrelated to the underlying pathology.

A 2021 report from the Alan Turing Institute, Britain's national institute for data science and AI, concluded that, of the hundreds of AI predictive tools developed, none were truly helpful and several were potentially harmful. One of the authors said that "This pandemic was a big test for AI and medicine. It would have gone a long way to getting the public on our side, but I don't think we passed that test."

Several problems were commonplace. For example, some algorithms used chest scans of healthy children for their training data for non-COVID cases; the algorithms learned to identify children's lungs, not COVID. Another training session used the healthy lungs of patients who were standing up and the lungs of seriously ill patients who were lying down. The algorithm focused on lung position. In other cases, the algorithms noticed that hospitals used different font styles to label the scans.

Another 2021 study was even more damning. An AI algorithm correctly identified the presence of COVID-19 even when the lung images were removed from the X-rays! The algorithm evidently noticed patterns in the outer borders of the images that happened to be correlated with the presence or absence of genuine COVID-19 pathology—which meant that the algorithm was useless for analyzing actual lung images.

The slow diffusion of AI in radiology hasn't diminished Geoffrey Hinton's deep optimism for deep learning. Undaunted by the setbacks, in November 2020, he said, "Deep learning is going to be able to do everything," a prediction that reminds me of Ed Yardeni's quip about the stock market: "If you give a number, don't give a date."

Is a Computer about to Take Your Job?

I began this chapter with Herbert Simon's 1965 prediction that "machines will be capable, within twenty years, of doing any work a man can do." Computers can certainly be useful assistants for doctors and baseball managers but they are not likely to replace them anytime soon. The same is true of most jobs.

When I was an undergraduate in the 1960s, I took a course devoted entirely to Robert Theobald's book *Free Men and Free Markets*. Theobald argued that technology was on the verge of creating an era of simultaneous abundance and unemployment as jobs were taken over by computers. He proposed that governments sever the link between work and wages by giving everyone a constitutional right to a guaranteed income. My term paper was deeply skeptical and graded harshly. I survived the low grade and remain skeptical.

More recently, we have Daniel Susskind's 2020 award-winning book, *A World Without Work*, and Elon Musk's glib comment: "What's going to happen is robots will be able to do everything better than us . . . all of us . . . When I say everything—the robots will be able to do everything, bar nothing." Really? Will robots soon be designing Teslas and making flippant predictions? AI's struggles recognizing a baseball bat and giving advice about walking downstairs backwards are not reassuring. I am reminded of the Danish proverb:

It is difficult to make predictions, especially about the future.

Efforts to predict which jobs are most at risk mainly reveal how much AI continues to overpromise and underdeliver. For example, a 2021 study by eminent professors at Princeton and NYU's Stern School of Business used a survey to collect yes/no responses for whether each of ten types of computer algorithms (such as image recognition, language translation, and video game programs) is related to each of 52 job-related abilities (including peripheral vision, finger dexterity, and memorization). A sample question: "Do you believe that image recognition by a computer or machine could be used for peripheral vision?" The relevant question, of course, is not whether it "could be used" but whether it would do a good job in the required context.

In addition, I tell my students that surveys with more than 10–20 questions are likely to be answered poorly or not at all. This jobs-at-risk survey had a mind-numbing 520 questions. Few people with any relevant expertise would be willing to spend the time required to give thoughtful answers. So, instead, the authors recruited thousands of low-paid gig workers from Amazon's Mechanical Turk web service. As the timeless saying goes: garbage in, garbage out.

The results, published in a top-tier strategic management journal, can best be described as laughable. They concluded that the occupations that are most at risk "consist almost entirely of white-collar occupations that require advanced degrees, such as genetic counselors, financial examiners, and actuaries." You may want to reread that sentence. Yes, they really did conclude that jobs requiring the most critical thinking skills are most at risk, even though critical thinking is the most glaring weakness of AI algorithms.

Thus, they concluded that judges are one of the most at-risk occupations while brickmasons, stone masons, and cafeteria attendants are among the safest. How could anyone, even ivory-tower professors, think that it is easier for an algorithm to be a judge than for a robot to lay bricks or serve cafeteria meals?

In case readers misunderstood their conclusion, the authors go into considerable detail explaining that surgeons are far more likely to be replaced by computers because surgeons have more cognative abilities:

Although the occupations require similar physical abilities, such as manual dexterity, finger dexterity, and arm-hand steadiness, ... a number of cognitive abilities related to problem solving, such as problem sensitivity, deductive and inductive reasoning, and information ordering, are highly important for surgeons but not for meat slaughterers. Our methodology suggests that occupations' exposure to AI largely stems from these cognitive abilities; accordingly, surgeons have a much higher [AI Occupational Exposure] than meat slaughterers.

The gig workers seemed to believe (and the professors evidently agreed) that computers are smarter than humans. Computers do have great memories, make fast calculations, and don't get tired, but that is not the intelligence required to be an excellent judge or surgeon.

An understandable conclusion is that jobs doing this kind of farcical research should be among the first jobs to be eliminated.

The Irony

Scientists created the computer algorithms that defeated the best human players in checkers, chess, Go, and *Jeopardy!* and fueled the idea that the grand promises of AI were on the verge of being realized. Scientists also created the recurring cycle of ambitious predictions, disappointing results, and moving goalposts that continue to be the norm.

False advertising ultimately undermines the credibility of the advertiser. The public has ample reasons for being disillusioned by the misleading claims that continue to be made for artificial intelligence. It's a hop, skip, and a jump from disillusionment to distrust.

CHAPTER 12

Artificial Unintelligence

In books, television shows, and movies, computers are like humans, but much smarter and less emotional. The less emotional part is right. Despite our inclination to anthropomorphize computers, computer algorithms do not have sentiment, feelings, or passions. They are also unintelligent, which makes the popular phrase *artificial intelligence* an oxymoron.

The Fire and Fuel of Thinking

In 1979, when he was just 34 years old, Douglas Hofstadter won a National Book Award and Pulitzer Prize for *Gödel, Escher, Bach: An Eternal Golden Braid*, which explored how our brains work and how computers might someday mimic human thought. He has now spent decades trying to solve this incredibly difficult puzzle. How do humans learn from experience? How do we understand the world we live in? Where do emotions come from? How do we make decisions? Can we write inflexible computer code that will mimic the mysteriously flexible human mind?

Hofstadter has concluded that analogy is "the fuel and fire of thinking." When humans see, hear, or read something, we can focus on the most salient features, the "skeletal essence," and understand this essence by drawing analogies. Hofstadter argues that human intelligence is fundamentally about collecting and categorizing human experiences, which can then be compared, contrasted, and combined. We create mental categories and then, when we see something new, it either reinforces a category we already have or we create a new category.

Perhaps when I was young I saw a slender piece of wood and was told that this is a *stick*. I created a mental category *stick* and every time I saw something similar and was told it is a stick, this reinforced my conception of what a stick is. It might have been a different type of wood, a different length or width, a different color, slightly bent, clean or dirty, but I still knew it was a stick. When I saw a baseball bat and was told that it is called a *baseball bat*, I created a new mental category. Baseball bats might be made of metal or wood, be different colors, or be different sizes, but I knew it was a bat if it had a characteristic shape and was intended to be used to hit a baseball.

I also created different mental categories for axes, cups, and bats with wings. In each case, I identified the essence of the object and I knew that the objects could come in different sizes, be made of different materials, or be different colors.

I also saw how baseball players held bats when they swung them at baseballs, I observed what happened to baseballs that were hit by bats; and I recognized that bats could be used to hit things other than baseballs. I knew that I should be careful when swinging a bat. I also learned how much bats cost and that they can break.

Computer algorithms do none of this. They live in Math World, devoid of any experiences in the real world, and this severely limits their capabilities.

Labeling without Understanding

For example, the essential difference between human observation and image-recognition software is that we can see things for what they are. The great physicist Richard Feynman famously explained the difference between knowing the name of something and understanding it:

[My father] taught me "See that bird? It's a brown-throated thrush, but in Germany it's called a halsenflugel, and in Chinese they call it a chung ling and even if you know all those names for it, you still know nothing about the bird—you only know something about people; what they call that bird. Now that thrush sings, and teaches its young to fly, and flies so many miles away during the summer across the country, and nobody knows how it finds its way," and so forth. There is a difference between the name of the thing and what goes on.

When humans see a wagon or a picture of a wagon, as in Figure 12.1, we recognize its skeletal essence—a rectangular box, wheels, and a handle. We know that a wagon will roll on its wheels when it is pushed or pulled. We know that wagons can carry things. We know that when a wagon is turned upside down, its contents will spill out. We know all this and more because we know what a wagon is and does—not just its name.

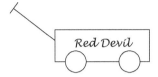

Figure 12.1 A simple wagon.

We know that wagons can be various sizes, made of various materials, or painted different colors, but we still recognize it as a wagon. We can look at it from many different angles, but we still recognize it as a wagon. It may be partly obscured by some other object, but we still recognize it as a wagon.

Not so with AI algorithms that input pixels and create mathematical representations of these pixels—what Turing winner Judea Pearl calls "just curve fitting." AI algorithms are trained by being shown many, many pictures of wagons, each with the label *wagon*. When an algorithm is shown a new picture, it curve-fits the pixels and looks for a mathematical match in its database. If it finds a good enough mathematical match with the wagon-labeled pixels it has been trained on, it returns the label *wagon* without knowing anything about wagons. If the new wagon picture is a different size, texture, or color; or viewed from a different angle; or partly obscured, the AI algorithm may flop.

When I tested the Clarifai deep neural network (DNN) image recognition program with the wagon image in Figure 12.1, it was 98 percent certain that the image was a business. The Wolfram Image Identification Project misidentified the wagon as a badminton racket. When the wagon color was changed to red, it was misidentified as a cigar cutter. When the red wagon was put on a 45° incline, it was misidentified as a paper clip.

OpenAI's CLIP Program

In 2015, Elon Musk and several other investors pledged a total of $1 billion to a non-profit artificial intelligence research company they named OpenAI, which would freely share its work with other companies in the hopes of developing "friendly AI." Musk resigned from the company's board of directors in 2018, but continued to donate money to the company. Shortly after Musk's resignation,

OpenAI converted to a for-profit company and received a $1 billion investment from Microsoft.

One of Open AI's products is a neural network algorithm called CLIP which it claims "efficiently learns visual concepts from natural language supervision." Traditionally, image classification programs match labels to images without making any attempt to understand what the label means. In fact, image categories are generally numbered because the label—like *wagon*—contains no useful information.

CLIP is said to be different because the program is trained on image–text pairs, with the idea being that the text will help the program identify an image. To employ the CLIP program, a user uploads an image and specifies proposed labels, at least one of which is correct. The program then shows its assessment of the probability of each label being correct.

I tried CLIP out with the wagon image in Figure 12.1 and the proposed labels *wagon*, *sign*, *badminton racket*, *goalposts*. Despite these helpful hints, CLIP favored *sign* (42 percent) and *goalposts* (42 percent), followed by *badminton racket* (9 percent) and *wagon* as the least likely label at 8 percent.

I wasn't sure if *goalposts* was the best spelling so I replaced it with *goal posts* and CLIP's probability of this being the correct label jumped from 42 percent to 80 percent! The sign label fell to 14 percent while *badminton racket* and *wagon* each came in at 3 percent. CLIP clearly did not really understand the text labels, since *goalposts* and *goal posts* are just alternate spellings of an identical concept.

CLIP is a bit of trickery in that the main purpose of the user-suggested labels is to help the algorithm by narrowing down the possible answers. This is not particularly useful since, in most real world applications, algorithms won't be fed helpful hints, one of which is guaranteed to be correct. As I tell my students, life is not a multiple-choice test.

More generally, the program's attraction to the labels *goalposts* and *goal posts* are convincing evidence that it is still just trying to match pixel patterns. No human would mistake this wagon image for goalposts.

The low probabilities assigned to *wagon* were also interesting. Perhaps the program is brittle because it was trained on red wagons and does not know what to make of a white wagon. I changed the color of the wagon to red and the *wagon* probability jumped to 77 percent when *goalposts* was spelled as one word and 69 percent when spelled as two words. Then I used a green wagon and the *wagon* probability dipped to 40 and 32 percent, depending on the *goalposts* spelling. Finally, for the red-and-white striped wagon in Figure 12.2, CLIP's *wagon* probabilities were 7 and 2 percent—which are even worse than for the white wagon. Offered a simple choice between wagon and candy cane, CLIP was 89 percent sure that the wagon in Figure 12.2 is a candy cane.

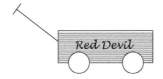

Figure 12.2 Wagon or candy cane?

Humans notice unusual colors, but our flexible, fluid mental categories can deal with color differences. For humans, the essence of a wagon is not its color. White wagons, red wagons, and red-and-white wagons are still wagons. Not so for image-recognition algorithms that curve-fit pixel patterns instead of recognizing and understanding entities. Living in Math World instead of the real world, they do not know what makes a wagon a wagon (the box, wheels, and handle) and cannot distinguish between important features and superficial details.

The label *artificial intelligence* is an unfortunate misnomer. The ways in which computers process data are not at all the ways in which humans interact with the world, and the conclusions computers draw do not involve the comprehension that is a crucial component of human intelligence. The label *artificial unintelligence* is more apt in that even the most advanced computer programs are artificial and unintelligent.

The Great America Novel Will Not be Written by a Computer

One of the claims OpenAI makes for its large language model GPT-3 is that, given a prompt, it can produce a remarkably coherent continuation of the prompt. Two skeptical NYU computer scientists, Gary Marcus and Ernest Davis, tested that claim. One of the prompts they tried was

You poured yourself a glass of cranberry juice, but then you absentmindedly poured about a teaspoon of grape juice into it. It looks okay. You try sniffing it, but you have a bad cold, so you can't smell anything. You are very thirsty. So

There are lots of reasonable ways this passage might be continued; for example,

you drink it anyway.

you take a sip, are pleasantly surprised, and drink the entire glass.

you take a sip and, not liking the taste, you pour yourself a fresh glass.

GPT-3 finished the prompt with

you drink it. You are now dead.

GPT-3 obviously doesn't know anything about cranberry juice or grape juice but, evidently, many passages in its database that contain words like "you can't smell anything. You are very thirsty" are followed by words like "you drink it. You are now dead." Analyzing statistical patterns in order to guess the next few words in a sentence can be an amusing parlor trick, but it is not indicative of understanding.

Marcus and Davis tried this prompt:

You are having a small dinner party. You want to serve dinner in the living room. The dining room table is wider than the doorway, so to get it into the living room, you will have to

An apt continuation might be something like these two possible answers:

turn the table sideways so that it fits through the door.
take the door off its hinges.

GPT-3's continuation was

remove the door. You have a table saw, so you cut the door in half and remove the top half.

If GPT-3 actually knew anything about the real world, it would know that sawing the door in half is a permanently bad idea and removing the top half of the door accomplishes nothing.

It is clear that GPT-3 does nothing more than string together words and phrases based on relationships among the words and phrases in its enormous database. As Marcus and Davis observe, GPT-3

learns correlations between words, and nothing more. It's a fluent spouter of bullshit, but even with 175 billion parameters and 450 gigabytes of input data, it's not a reliable interpreter of the world.

A final example. Marcus and Davis used this prompt:

At the party, I poured myself a glass of lemonade, but it turned out to be too sour, so I added a little sugar. I didn't see a spoon handy, so I stirred it with a cigarette. But that turned out to be a bad idea because

GPT-3 continued with this inexplicable nonsense:

it kept falling on the floor. That's when he decided to start the Cremation Association Here, the same type face was used for the question and answer of North America, which has become a major cremation provider with 145 locations.

Unable to understand words, algorithms struggle to distinguish between fact and fiction. GPT-3 recycles all sorts of unfiltered garbage it finds on the Internet, including allegations that Bill Gates is responsible for COVID-19 and that vaccines are a scam.

In one notorious 2017 example, Google Home One was asked, "Is Obama planning a coup?" It responded,

According to details exposed in Western Center for Journalism's exclusive video, not only could Obama be in bed with the communist Chinese, but Obama may in fact be planning a communist coup d'etat at the end of his term in 2016!

Never mind that 2016 had already passed without incident, Google's algorithm was just parroting a bogus story from a conspiracy news site because it has no reliable means of assessing the reliability of the terabytes of text it processes. This is a problem not just for screening conspiracy theories but for the evaluation of health advice, medical diagnoses, and all sorts of prediction models. As long as algorithms have no effective way of distinguishing trustworthy sources from unreliable garbage, the algorithms themselves will be suspect.

In 2021, on the day after Christmas, a restless 10 year old asked Amazon's Alexa for a challenge and Alexa responded, "The challenge is simple. Plug in a phone charger about halfway into a wall outlet, then touch a penny to the exposed prongs." Fortunately, the child's mother was in the room at the time and yelled, "No, Alexa, no!," like scolding a misbehaving dog, while making sure that her child did not try the challenge.

A few months earlier, a Google user reported that if you queried "had a seizure now what" Google's search engine returned a list of suggested actions that was actually a list of things *not to do* that Google's search engine had found but obviously not understood.

BABEL

College and university admission decisions are increasingly dependent on Automated Essay Scoring (AES) algorithms, including the Educational Testing

Service (ETS) e-rater and Vantage Learning's Intellimetric. The ETS tells students that "The primary emphasis in scoring the Analytical Writing section is on your critical thinking and analytical writing skills rather than on grammar and Mechanics." The truth is exactly the opposite. AES algorithms are utterly incapable of assessing critical thinking and analytical writing skills.

Because the algorithms do not understand what words mean, they necessarily rely on quantifiable measures such as the use of uncommon words and the length of words, sentences, paragraphs, and essays. Students can consequently use large uncommon words in long sentences in babbling paragraphs in tiresome essays to game the system.

Les Perelman, former director of Writing Across the Curriculum at MIT, worked with three undergraduate students (two from MIT, one from Harvard) once a week for four weeks to develop Basic Automatic BS Essay Language Generator (BABEL) to demonstrate the inadequacies of AES.

Here is a snippet from a BABEL 192-word, 13-sentence paragraph:

Theatre on proclamations will always be an experience of human life. Humankind will always encompass money; some for probes and others with the taunt. Money which seethes lies in the realm of philosophy along with the study of semiotics. Instead of yielding, theatre constitutes both a generous atelier and a scrofulous contradiction.

The writing is chock full of unusual words, yet utterly meaningless. E-rater has no way of assessing whether passages are accurate and coherent, let alone persuasive, and gave it the highest possible score (6 on a scale of 1 to 6).

Michael Shermis, an AES enthusiast, reported that Abraham Lincoln's Gettysburg Address received scores of only 2 or 3 from the several AES systems he tested. Maybe the algorithms didn't like the fact that Lincoln's 278 words were broken into three paragraphs—one consisting of a single sentence. Or maybe there were not enough unusual words. What is certain is that the content did not matter a whit to the algorithms.

ETS's response has been to modify e-rater so that it might detect and flag BABEL-generated essays. As Perelman remarked, this is a "solution in search of a problem." Test-takers are not going to use BABEl when they take these tests. The point of the BABEL experiment was to demonstrate that grading algorithms are inherently flawed in that they do not assess the primary goals of writing, such as conveying information and persuading readers.

Alert students can game the system by writing gaseous prose filled with obscure words. Alert instructors can teach their students to game the system. The consequence of algorithmically graded tests of writing ability will be to degrade writing. In Perelman's words, students are being taught to write "verbose and pretentious prose."

The AI Marketing Ploy

Justin Wang received a bachelor's degree from Murdoch University in 2012 and a master's degree in Information Technology Management from the University of Sydney in 2016. In January 2017, he founded a Singapore-based company with the mysteriously cool name Scry in order to "manage information technology to achieve business goals, as well as—and perhaps more importantly—how it can be wielded to disrupt existing value networks."

What's behind the mystery and jargon? It turns out that Scry is a "social forecasting platform." Users join for free and can enter their personal estimates of the probabilities that certain events will happen, with Scry calculating the average probability. For example, one question is "Will Apple launch a commercial self-driving electric vehicle before the end of 2024?" As I write this, there are 18 responses, entered up to six months ago. Eight answers are 50–50 and two are 100 percent *yes*. Overall, the average probability is 57 percent. This seems like harmless fun, inspired by the fickle wisdom of crowds. One glaring weakness is that users see the probabilities others have chosen (and the average value) before they enter their own probabilities. The wisdom-of-crowds argument assumes that individual assessments are made independently. That assumption is sometimes untrue and is surely false here.

Soon, Wang decided to monetize these predictions by creating Scry Vantage:

We turn any questions you have about the future into measurable and accurate predictions, made through our proprietary AI.

No details are given beyond the price: "starting at $400/prediction." I suspect that if a question is already in the Scry database, then the user-provided probabilities are the basis of the $400+ answer, but I wasn't about to spend $400+ to confirm this. In any case, the calculation of an average probability is hardly AI.

If a user pays $400+ and asks a new question, then the question might be forwarded to Scry users, although it would take an embarrassing long time to get an answer. Perhaps Wang makes a semi-educated guess? Maybe he asks a handful of friends to make semi-educated guesses? Maybe he asks GPT-3?

I asked GPT-3 five times and, oddly enough, got exactly the same evasive answer every time:

GARY: What is the probability that Apple will launch a commercial self-driving electric vehicle before the end of 2024?

GPT-3: It is highly unlikely that Apple will launch a commercial self-driving electric vehicle with complete pricing and performance information for the public before the end of 2024. Based on my current understanding, it is impossible for Apple to complete the development of a commercially viable self-driving electric vehicle on its own before that date. Apple's competitor, Google, has a project in the works but there are still some key challenges that need to be addressed in order for that project to materialize in a saleable product before the end of 2024.

Scry and GPT-3 do not use an AI algorithm to analyze data and calculate probabilities. Searching the Internet and elsewhere for opinions is qualitatively different from using an empirical model to estimate the probability that a certain company will make a certain product by a certain date.

Understandably, the Scry website gives absolutely no information about their purported "proprietary AI." The attitude seems to be "We call it AI; that's all you need to know." The word "AI" was selected by the Association of National Advertisers as the Marketing Word of the Year in 2017 and, too often, it seems that AI has become just a marketing ploy.

Scry is a small, but clear, example of the presumption by too many businesses that people will be impressed by anything that advertises itself as AI. Larger scale examples are OpenAI's CLIP image-recognition algorithm and OpenAI's GPT-3 text-generating algorithm which purportedly use an understanding of words to identify images and write persuasive prose, but don't.

I am reminded of Long-Term Capital Management, a hedge fund launched in 1994 by Solomon Brothers superstar John Meriwether who had put together a dream team that included Myron Scholes and Robert C. Merton (who would win Nobel prizes in 1997); several MIT PhDs who had worked for Meriwether at Solomon; and David Mullins, another MIT PhD, who left his position as vice-chairman of the Federal Reserve Board, where he was expected to succeed Alan Greenspan as Fed Chair. What could go wrong?

The minimum investment was $10 million and the only thing investors were told about Long-Term's strategy was that the management fees would be 2 percent of assets plus 25 percent of profits. A legendary portfolio manager ("Dave") told me that he had been pitched by Long-Term but chose not to invest because the only thing he knew for certain was that they were greedy. Other investors were not so cautious. Long-Term raised more than $1 billion.

After a few stellar years, Long-Term crashed in 1998. Meriwether promptly launched a new hedge fund, named JWM Partners. JWM crashed in 2009, and Meriwether started yet another hedge fund called JM Advisors Management.

Fool me once, shame on you. Fool me twice, shame on me. Fool me three times, you have no shame.

Warren Buffett has often warned, "Never invest in a business you don't understand." Dave heeded that advice when Long-Term came looking for his money. We should all heed that advice when a person or company says that they are using AI and doesn't give a clear explanation of exactly what that means. Don't trust an AI algorithm you don't understand.

Making Lemons out of Lemonade

An insurance company with the quirky name Lemonade was founded in 2015 and went public in 2020. In addition to raising hundreds of millions of dollars from eager investors, Lemonade quickly attracted more than a million customers with the premise that AI algorithms can estimate risks accurately and that buying insurance and filing claims can be fun:

Lemonade is built on a digital substrata—we use bots and machine learning to make insurance instant, seamless, and delightful.

Adding to the delight are the friendly names of their bots, like AI Maya, AI Jim, and AI Cooper.

The company doesn't explain how its AI works, but there is this head-scratching boast:

A typical homeowners policy form has 20-40 fields (name, address, bday . . .), so traditional insurers collect 20-40 data points per user. AI Maya asks just 13 Q's but collects over 1,600 data points, producing nuanced profiles of our users and remarkably predictive insights.

This mysterious claim is, frankly, a bit creepy. How do they get 1600 data points from 13 questions? Is their app using our phones and computers to track everywhere we go and everything we do? The company says that it collects data from every customer interaction but, unless it is collecting trivia, that hardly amounts to 1600 data points.

How do they know that their algorithm is "remarkably predictive" if they have only been in business for a few years? Lemonade's CEO and co-founder

Daniel Schreiber has alluded to the fact that "AI crushes humans at chess, for example, because it uses algorithms that no human could create, and none fully understand." In the same way, "Algorithms we can't understand can make insurance fairer."

An example he gives is not reassuring.

Let's say I am Jewish (I am), and that part of my tradition involves lighting a bunch of candles throughout the year (it does). In our home we light candles every Friday night, every holiday eve, and we'll burn through about two hundred candles over the 8 nights of Hanukkah. It would not be surprising if I, and others like me, represented a higher risk of fire than the national average. So, if the AI charges Jews, on average, more than non-Jews for fire insurance, is that unfairly discriminatory?

His answer:

It would definitely be a problem if being Jewish, per se, resulted in higher premiums whether or not you're the candle-lighting kind of Jew. Not all Jews are avid candle lighters, and an algorithm that treats all Jews like the "average Jew," would be despicable.

So far, so good. His solution:

[An] algorithm that identifies people's proclivity for candle lighting, and charges them more for the risk that this penchant actually represents, is entirely fair. The fact that such a fondness for candles is unevenly distributed in the population, and more highly concentrated among Jews, means that, on average, Jews will pay more. It does not mean that people are charged more for being Jewish.

Schreiber says that this is "a future we should embrace and prepare for" because it is "largely inevitable . . . .Those who fail to embrace the precision underwriting and pricing . . . will ultimately be adversely-selected out of business."

I don't know if this future is inevitable, but I am withholding my embrace. An algorithm might be able to identify a reasonably accurate proxy for having at least one Jew in a household but if the algorithm takes this Jewishness into account, it is discriminatory. Since algorithms can't currently identify candle-lighting proclivities, what can they use other than proxies for being Jewish? Since Lemonade is using a black-box algorithm that "we can't understand," it may well be discriminatory—and we have no way of knowing for certain.

Looking ahead to Schreiber's predicted inevitable future, how would an algorithm move beyond Jewish proxies to collecting data on a household's fondness for candle lighting? Would it use customer smartphone cameras to record what goes on inside their homes? Would it ransack customer credit card statements for evidence of candle-buying—which might lead people to pay cash for candles the same way that some people pay cash for illegal drugs?

We are mired in a despicable place where Lemonade's black box algorithm may well be discriminatory—not just against Jews—and may be seeking to bulldoze our privacy.

In May 2021 a very specific problem arose when Lemonade tweeted that

When a user files a claim, they record a video on their phone and explain what happened. Our AI carefully analyzes these videos for signs of fraud. [AI Jim] can pick up non-verbal cues that traditional insurers can't, since they don't use a digital claims process. This ultimately helps us lower our loss ratios (aka how much we pay out in claims vs. how much we take in).

Are claims really being validated by non-verbal cues (like the color of a person's skin) that are being processed by black-box AI algorithms that the company admits it does not understand?

There was a media uproar since AI algorithms for analyzing people's faces and emotions are notoriously unreliable and biased. Lemonade had to backtrack. A spokesperson said that Lemonade was only using facial-recognition software for identifying people who file multiple claims using multiple names. But if Lemonade is using image-processing software, there is no way of knowing what their black-box algorithm is doing with these data.

Lemonade then tried to divert attention from image-processing software by claiming that AI Jim is not really AI, but just an algorithm for recording customer information that is checked against preset rules.

It's no secret that we automate claim handling. But the decline and approve actions are not done by AI, as stated in the blog post. [Lemonade will] never let AI, in terms of our artificial intelligence, determine whether to auto reject a claim. We will let AI Jim, the chatbot you're speaking with, reject that based on rules.

The lemonade is smelling a little sour at this point. In a pre-IPO filing with the SEC, Lemonade stated that "in approximately a third of cases [AI Jim] can manage the entire claim through resolution without any human involvement." Lemonade has also boasted that AI Jim uses "18 anti-fraud algorithms" to assess claims. Are these 18 algorithms just checkboxes?

Overall, it makes sense that an insurance company can pare costs by having fewer sales agents and office buildings. However, it seems a stretch to say that buying insurance and filing claims can be delightful. Insurance purchases should involve some thoughtful consideration of the coverage, deductibles, price, and so on, not light-headed giddiness. Nor are people likely to be delighted by a goofy app after they have been in an automobile accident, had their home burn down, or suffered any other substantial loss that warrants a claim.

I would be pleased—not delighted—if the process were simple and relatively painless. Lemonade deserves credit for trying really hard to do this. On the other hand, as for AI, Lemonade seems to fit a common pattern—put an AI label on a business and hope that investors and customers are impressed.

Much of what the Lemonade bots do is apparently just helping customers walk through routine forms. If the bots really are collecting 1600 data points from each customer interaction and analyzing these data with a black-box data-mining algorithm, then all of the many cautions about data mining apply. Specifically, its algorithms may well be discriminatory and their boast that that their algorithms are "remarkably predictive" are most likely based on how well they predict the past—which is a fundamentally unreliable guide to how well they predict the future.

Lemonade's losses have grown every quarter and, as I write this on June 14, 2022, its stock price is $16.65, down more than 90% from its high of $183.26 on January 11, 2021.

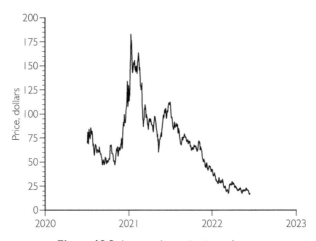

Figure 12.3 Lemonade turning into a lemon

The Irony

Computers are terrific. I use them every day—not just for writing books, research papers, and op-eds but for mathematical calculations, statistical analyses, and simulations that would otherwise be essentially impossible. I have three grown children who are computer/data scientists and could not do what they do without computers.

I also recognize that, as wondrous as they are, computers are not intelligent. Attempts to treat them as intelligent are bound to frustrate. Attempts to market them as intelligent are bound to disappoint. Those who are disillusioned by the hype are rightfully skeptical and distrustful.

PART V

The Crisis

CHAPTER 13

Irreproducible Research

Many dodgy studies have been exposed as dodgy when tested with fresh data; for example, claims that hurricanes are more dangerous if they have female names, that Asian-Americans are susceptible to heart attacks on the fourth day of the month, and that people postpone death so that they can celebrate the Harvest Moon Festival, Passover, birthdays, and other important events.

An even more fundamental problem is that some published research can't be confirmed with the original data, let alone corroborated with fresh data. Thus, a distinction can be made between *reproducibility* (whether others using the original data obtain the reported results) and *replication* (whether others using fresh data obtain results consistent with those originally reported).

I will talk about reproducibility issues in this chapter, replication issues in the next. Together, these constitute an enormous scandal that undermines public faith in the reliability of science and provides fuel for those who want to politicize and delegitimize science.

Incorrect p-Values

Most empirical papers test hypotheses by using p-values that are determined by the means, standard deviations, and other statistics. Since the test statistics and p-values are standard output from thoroughly tested software packages, we expect the values reported in published papers to be accurate. However, a team of statisticians checked whether the p-values in more than 250,000 papers published in eight flagship psychology journals between 1985 and 2013 matched the test statistics reported in the papers and found that half of the articles had at least one incorrect p-value and 13 percent of the articles had a gross error in which the p-value was reported to be less than 0.05 but the true p-value was not, or vice versa. The gross errors tended to support the authors' expectations, suggesting that the mistakes were not accidental. Other studies, from a variety of fields, have come to similar conclusions.

Making Data Publicly Available

These p-value double-checks assume that the reported test statistics behind the p-values are correct—which is difficult to determine without access to the original data. Historically, researchers seldom reported their raw data—in part because journal space was precious. This argument is no longer compelling since online appendixes are essentially free.

Many journals now require authors to make any nonconfidential data available to others. Unfortunately, if the data are not posted online before the article is published, there is little that can be done to enforce such agreements. I do a lot of reanalyses of papers that make suspect claims and I am often rebuffed in my requests to obtain the data on which the claims are based. My experience is hardly unique.

Four psychologists from the University of Amsterdam contacted the authors of 141 empirical papers published in four major American Psychological Association journals. All authors had signed an APA Certification of Compliance that includes a clause on "Sharing Research Data for Verification":

After research results are published, psychologists do not withhold the data on which their conclusions are based from other competent professionals who seek to verify the substantive claims through reanalysis and who intend to use such data only for that purpose, provided that the confidentiality of the participants can be protected and unless legal rights concerning proprietary data preclude their release.

Nonetheless, only thirty-eight authors (27 percent) shared their data.

Since 2011, *Science*, one of the world's top research journals, has required all authors to make their data and computer code freely available. The requirement is clear and unambiguous:

All data necessary to understand, assess, and extend the conclusions of the manuscript must be available to any reader of *Science*. All computer codes involved in the creation or analysis of data must also be available to any reader of *Science*.

Yet a 2018 analysis of 204 papers published in *Science* after the implementation of this policy found that only 44 percent of the authors shared enough information to allow an attempt to reproduce the results and only 26 percent of the results could be reproduced.

Some of the responses to the data requests reflected an apparent hostility to making data publicly available:

> When you approach a PI for the source codes and raw data, you better explain who you are, whom you work for, why you need the data and what you are going to do with it.

> The data files remains our property and are not deposited for free access.

> We do not typically share our internal data or code with people outside our collaboration.

Scientific Sleuths

The stonewalling by some authors has energized a small group of data detectives. One is Nick Brown, who was involved in the pizza papers scrutiny described in Chapter 5. At the age of fifty and after a recent career change from IT to human resources, Brown happened to talk to psychologist Richard Wiseman about what Brown saw as the lack of evidence for many learning and development tools that he was responsible for purchasing. Curiosity piqued, Brown decided to get a master's degree in psychology at the University of East London. The assigned readings in one class included a famous paper which claimed that a 2.9013 ratio of positive to negative emotions separated individuals, marriages, and business teams that flourished from those that languished. One of the authors, Marcial Losada, immodestly named this magic ratio "the Losada line."

In retrospect, it is astonishing that anyone—let alone psychologists—would believe that a precise number could be applied universally to something as amorphous as human emotions. The paper used physics and engineering equations that describe the movement of fluids, presumably to bolster the credibility of the argument by making it appear scientific but, in reality, making the argument even sillier. Losada's borrowing of the language of fluid dynamics might easily be mistaken for a parody: high-performance teams operate "in a buoyant atmosphere created by the expansive emotional space;"

low-performance teams are "stuck in a viscous atmosphere highly resistant to flow."

Brown knew instinctively that this was BS, but he wasn't confident about his knowledge of math or psychology; so he teamed up with Alan Sokal, a well-known physics professor and BS debunker, and Harris Friedman, a distinguished psychology professor, to co-author a devastating critique. A brief excerpt:

We find no theoretical or empirical justification for the use of differential equations drawn from fluid dynamics, a subfield of physics, to describe changes in human emotions over time; furthermore, we demonstrate that the purported application of these equations contains numerous fundamental conceptual and mathematical errors. The lack of relevance of these equations and their incorrect application lead us to conclude that Fredrickson and Losada's claim to have demonstrated the existence of a critical minimum positivity ratio of 2.9013 is entirely unfounded.

Another data detective, James Heathers, now in his 30s, grew up in Australia and earned bachelor's, masters, and doctorate degrees at the University of Sydney. Brown and Heathers are an odd couple. Brown is quiet and cautious; Heathers is loud and boisterous. Brown says modestly that he is a "self-appointed data police cadet." His blogging site has no photograph and essentially no personal information. Heathers calls himself a "data thug" and goes out of his way to bolster that image. His LinkedIn page features a menacing photo with strands of hair crossing his nose, looking like, well, a data thug.

They met online when Brown was looking for some background information on the vagus nerve for an analysis of a paper by Barbara Fredrickson, the lead author of the "Losada line" paper; Heathers ended up as lead author on their subsequent rebuttal article.

As invaluable as scientific sleuthing is, it is not a popular occupation. As with doctors, the police, and many other occupations, there seems to be an unwritten code of silence among scientific researchers. Heathers has written that in academia "the squeaky wheel gets Tasered."

One consequence is that researchers who suspect sloppy science but don't want to make enemies send their suspicions to Brown, Heathers, and other thick-skinned science sleuths who, I imagine, know they are doing noble work and enjoy the challenge, like solving a difficult Sudoku puzzle.

Hairstyles and High Heels

In 2015 Brown saw a tweet by a British science blogger who goes by the moniker of Neuroskeptic, about a study that concluded that men are more likely to help women whose hair hangs loose. The study was by a French psychologist named Nicolas Guéguen who is semi-famous or fully notorious, depending on your point of view, for studies claiming that men are more likely to approach a woman drinking alone in a bar if she is wearing high heels; male diners give larger tips to waitresses with blonde hair; and women are more likely to give men their telephone numbers if the sun is shining. Guéguen has also reported that men are more likely to approach large-breasted women in night clubs and that male drivers are more likely to give lifts to a female hitchhiker if the color of the t-shirt she is wearing is red or if she has large breasts.

Another study had volunteers approach members of the opposite sex and ask, "Will you come to my apartment to have a drink?" or "Would you go to bed with me?" Based on the responses, Guéguen concluded that "men are apparently more eager for sexual activity than women are." These are the kinds of studies that get worldwide attention but also make you wonder why the researchers are not doing something more useful with their lives.

Brown mentioned the loose-hair study in passing during a Skype conversation with Heathers and, after the laughter subsided, they decided to do a deep dive into the data. Guéguen had recruited a 19-year-old woman who dressed in black and wore her hair either loose, tied in a ponytail, or twisted in a bun. As she walked past a male or female pedestrian "aged roughly between 25 and 45," she took her hands out of her coat pockets and dropped a glove. The pedestrian was given a score of 3 for picking up the glove and handing it to the woman, a score of 2 for telling her that she had dropped a glove, and a score of 1 for doing nothing. This scenario was acted out with ninety male and ninety female pedestrians.

The results are shown in Table 13.1. The differences in male helpfulness scores are statistically significant, with men far more helpful when the woman had a natural hair style. For female pedestrians, the hairstyle differences are not statistically significant.

The most obvious oddity is the unrealistically large reported effect of hair style on male behavior. Brown and Heathers also noticed that the second decimal point for all six of the reported average scores (highlighted in bold) is 0. Perhaps the averages (or the underlying data) were made up?

Table 13.1 *Average Helpfulness Scores*

	Hairstyle			
	Natural	Ponytail	Bun	Total
Male pedestrians	**2.80** (0.41)	**1.80** (0.76)	**1.80** (0.76)	2.13
Female pedestrians	**1.80** (0.41)	**1.60** (0.81)	**1.60** (0.50)	1.67
Total	2.30	1.70	1.70	1.90

Note: standard deviations are in parentheses.

Brown and Heathers have a neat little algorithm that identifies whether the reported means and standard deviations are possible when the data are integers (like 1, 2, and 3). When they applied their algorithm to the six means and standard deviations in Table 13.1 they found that, in all six cases, each score of 1, 2, and 3 had to appear exactly 6, 12, 18, or 24 times. Consider, for example, the female pedestrians with the volunteer wearing her hair in a bun. A mean of 1.60 and standard deviation of 0.50 is only possible if the thirty scores consist of twelve 1s and eighteen 2s. The chances that each of the 1, 2, 3 scores would appear exactly 6, 12, 18, or 24 times in all six cases are somewhere between slim and none.

Brown and Heathers figured that, in the same way that many burglars are repeat offenders, some of Guéguen's other titillating papers might be suspect. Sure enough, they found similar oddities in nine other Guéguen papers— implausibly large effects and improbable data. There were other problems. In one study, male volunteers asked 500 women who looked like they were between the ages of 18 and 25 for their telephone numbers. Incredibly, Guéguen reported that when the women were asked their ages afterward, every woman answered (even the women who refused to give their telephone numbers) and every woman was, indeed, between 18 and 25.

After years of ignoring or dodging requests from Brown and Heathers and the French Psychological Society (SFP) for raw data and explanations of the suspicious quirks in his reported results, Guéguen was investigated in 2018 by the Research Integrity Officer of Rennes 2 University, which concluded that there had been a number of questionable research practices. These breaches of scientific integrity involved data, statistical calculations, and research methodologies. Nicolas Guéguen revealed, in particular, that he often based his papers on fieldwork done by undergraduate students, though he neglected to credit them in his papers. In November 2018, Brown and Heathers were contacted by a student who had taken a class of Guéguen's (quotation translated from French by Brown and Heathers):

As part of an introductory class entitled "Methodology of the social sciences," we had to carry out a field study.... [M]ost of the students were not very interested in this class. Plus, we were fresh out of high school, and most of us knew nothing about statistics. Because we worked without any supervision, yet the class was graded, many students simply invented their data. I can state formally that I personally fabricated an entire experiment, and I know that many others did so too.... At no point did Dr. Guéguen suggest to us that our results might be published.

To date, only two of Guéguen's papers have been retracted—the one reporting that women wearing high heels are more attractive to men and another reporting that men carrying guitars are more attractive to women—though Heathers and Brown are convinced that many more papers should be retracted. One bit of good news is that Guéguen seems to have essentially stopped writing papers, and may have found something better to do.

The Ivermectin Silver Bullet

In August 2021, it was reported that the chairman of the Tokyo Metropolitan Medical Association had proclaimed that "Now is the time to use ivermectin" for treating COVID-19. In October a prominent U.S. conspiracy promoter said on his radio blog that "Japan has PULLED the vaccines and substituted Ivermectin—and in one month, wiped COVID out in that country!", a claim echoed and amplified over the Internet. One site dedicated to "defending health, life and liberty" ran a story headlined "Japan ends vaccine-induced pandemic by legalizing IVERMECTIN, while pharma-controlled media pretends masks and vaccines were the savior."

Ivermectin is an anti-parasitic drug that had not been approved for the treatment of COVID-19 in the United States or Japan, nor has Japan stopped using vaccines. In fact, the rapid decline in cases is due to the success of Japan's vaccination campaign. Nonetheless, ivermectin may be effective and it has been recommended by health officials in a half-dozen countries.

One reason for the optimism was the November 2020 report of a large clinical trial conducted by doctors affiliated with Benha University in Egypt. Among 600 patients with severe COVID-19, some were given ivermectin while those in the control group were given hydroxychloroquine, another suspect cure. (It would have been better to give those in the control group a placebo; otherwise we won't know how much of the observed difference in results might be due to the positive effects—if any—of ivermectin and how much might be due to the negative effects of hydroxychloroquine.)

The researchers reported an astonishing 90 percent reduction in deaths in the ivermectin group. Soon, a U.S. doctor who had worked at the University of Wisconsin Health University Hospital was testifying before Congress that ivermectin was a "miracle drug" with "Nobel prize-worthy" effects.

Others were skeptical. Gideon Meyerowitz-Katz, self-described as Health Nerd, wrote that the 90-percent cure-rate, "if true would make ivermectin the most incredibly effective treatment ever to be discovered in modern medicine."

A British biomedical master's student named Jack Lawrence read the Egyptian paper for one of his courses and immediately noticed several issues, including the fact that the Introduction had been plagiarized. He contacted Nick Brown, who was becoming well known as a scientific sleuth.

Brown found page after page of problems. For instance, the study reported recovery times of 9 to 25 days with a mean of 18 days and a standard deviation of 8 days. Those numbers could only be accurate if most patients recovered in either exactly 9 days or exactly 25 days. That was possible, but not very likely. Even worse, Kyle Sheldrick, a Sydney doctor and researcher, found several standard deviations that were "mathematically impossible."

The underlying data were said to be available online, but it turned out to cost $9 plus tax per month and, even after paying $9 plus tax, an unknown password was required. Lawrence paid the $9 plus tax and guessed the password (1234).

The raw data turned out to be a mess. Sheldrick noted that there were absolutely no missing data on ages, birthdates, or anything else. In these kinds of studies, there are inevitably missing data—unless the researchers made up the numbers.

There were, in fact, several facets of the data that suggested the data were fabricated—too many to list here, but I will mention a few. The report said that the study began on June 8, 2020, but the data file indicated that a third of the people who died from COVID-19 died before June 8. In addition, the data for several patients match the data for other patients closely, as if someone had cut-and-pasted data and then tried to disguise these cloned patients by changing a few digits or letters. The ages of 410 of the 600 patients are even numbers, compared to 190 odd-numbered ages. The probability that large a disparity would happen by chance is less than 1 in 10 quintillion.

In addition, a large number of statistical calculations reported in the paper do not match the raw data. The authors reported that they used a statistical package called SPSS but they also reported that "After the calculation of each of the test statistics, the corresponding distribution tables were counseled [*sic*] to get the 'P' (probability value)." Decades ago, researchers looked up p-values in tables. Nowadays, SPSS and other statistical packages display the p-values right next to the test statistics. There is no need to consult probability tables. Did the authors actually use SPSS?

The study was eventually retracted, as was a Lebanese study with similar data problems. The scientific sleuths found problems with several other ivermectin studies, including one in which the researchers, in Heathers' words, "said they recruited participants from hospitals that had no record of having participated in the research, and then blamed mistakes on a statistician who claimed never to have been consulted."

The results of more reliable studies are now in and have not found any benefits. The American Medical Association, American Pharmacists Association, Food and Drug Administration, U.S. Centers for Disease Control, and World Health Organization have all warned that ivermectin is not approved or recommended for preventing or treating COVID-19. Merck, the pharmaceutical company that developed ivermectin, issued a statement saying that there is "No scientific basis for a potential therapeutic effect against COVID-19 . . . ; no meaningful evidence for clinical activity or clinical efficacy in patients with COVID-19 disease, and a concerning lack of safety data."

Publish or Perish

The linchpin of scientific progress is that scientists publish their findings so that others can learn from them and expand on their insights. This is why some books are rightly considered among the most influential mathematical and scientific books of all time:

Elements, Euclid, *c.*300 BC

Physics, Aristotle, *c.*330 BC

Dialogue Concerning the Two Chief World Systems, Galileo Galilei, 1632

Mathematical Principles of Natural Philosophy, Isaac Newton, 1687

On the Origin of Species, Charles Darwin, 1859

As Newton said, "If I have seen further it is by standing on the shoulders of Giants."

It seems logical to gauge the importance of modern day researchers by how many papers and books they have published and how often their work has been cited by other researchers. Thus, the cruel slogan "publish-or-perish" has become a brutal fact of life. Promotions, funding, and fame all hinge on publications demonstrating that a researcher is worthy of being promoted, supported, and celebrated. Anirban Maitra, a physician and scientific director at MD Anderson Cancer Center, wryly observed that "Everyone recognizes it's a hamster-in-a-wheel situation, and we are all hamsters."

Every job interview, every promotion case, every grant application includes a publication list. Virtually every researcher has a website with a curriculum vitae (CV) link. The Internet now facilitates tabulations of how often an author's papers are cited by other researchers.

In 2004 Google launched Google Scholar, a database consisting of hundreds of millions of academic papers found by its web crawlers. Google then created Scholar Citations, which lists a researcher's papers and the number of citations found by the Googlebots. Soon, Google began compiling productivity indexes based on the number of papers an author has written and the number of times these papers have been cited. For example, the h-index (named after its creator Jorge E. Hirsch) is equal to the number of articles h that have been cited at least h times. My h-index is currently 25, meaning that I have written twenty-five papers that have each been cited by at least twenty-five other papers.

Beyond helping researchers find relevant research done by others, Google Scholar is a fast-and-easy (okay, lazy) way of assessing productivity and importance. As Google says, its citation indexes are intended to measure "visibility and influence." Citation indexes are now considered so important that some researchers include them on their CVs and webpages.

Unfortunately, the publish-or-perish culture has encouraged researchers to game the system, which undermines the usefulness of publication and citation counts. This is an example of Goodhart's law, first mentioned in Chapter 5: "When a measure becomes a target, it ceases to be a good measure." For publish-or-perish, the widespread adoption of publication counts and citation indexes to gauge visibility and influence has caused publication counts and citation indexes to cease to be good measures of visibility and influence.

The first sign of this breakdown was an explosive growth in the number of journals in response to researchers' ravenous demand for publication numbers. In 2018 it was estimated that more than 3 million articles were published in more than 42,000 peer-reviewed academic journals. It has been estimated that half of all peer-reviewed articles are not read by anyone other than the author, journal editor, and journal reviewers, though I cannot think of any feasible way of identifying articles that no one has read. It is nonetheless surely true that many articles are read by very few, but the publish-or-perish incentive structure makes it better to publish something that no one reads than to publish nothing at all.

Another way to pump up citation counts is hyper-authorship—papers with literally hundreds or even thousands of co-authors. Add my name to your list of co-authors and I will add your name to my list. Between 2014 and 2018 there were 1,315 papers published with 1,000 or more co-authors. A 2004 paper had 2,500 co-authors, a 2008 paper had 3,100; and a 2015 paper had 5,154 (it took twenty-four of the paper's thirty-three pages to list the co-authors). A 2021 paper set the Guinness record with 15,025 co-authors and, unsurprisingly, celebrated this dubious achievement on Twitter.

Gaming Peer-Review

Citation counts also encourage a variety of unseemly tricks to pass peer-review and accumulate citations. When a researcher (or team of researchers) finishes a study, the results can be written up and submitted to an academic journal. A journal editor gives the paper a quick once-over and decides whether it is worth sending to experts for a full peer-review. After reading the referee reports, the editor decides whether to either reject the paper or send the author(s) a revise-and-resubmit request, after which the revised paper will be resent to the reviewers for possible acceptance. On rare occasions, a paper might be accepted with no revisions.

In theory, the peer-review process enhances the reliability of published work by ensuring that papers do not get published unless impartial experts in the field deem the papers worthy of publication. Peer-review is well-intentioned, but flawed in many ways.

First, the authors of empirical papers have traditionally not been expected to share their data. Even if they do, reviewers have not been expected to examine the data closely and check the calculations. Second, the best researchers are incredibly busy and naturally more inclined to do their own research than to review someone else's work. Those generous souls who do agree to review papers may be motivated by an altruistic loyalty to their profession and/or duty to serve as gatekeepers.

In practice, peer-review is often cursory or else done by people who have ample time on their hands because they do relatively little research (and are consequently not well-informed about the current state of the field). Ill-informed and contradictory reviews are familiar to everyone who publishes in academic journals.

From the publishers' perspective, the publish-or-perish pressure on researchers is an opportunity. Instead of publishing one journal in a field, a publisher might create multiple specialty journals. Instead of publishing four times a year, a journal might be published six or twelve times a year. Unlike books where authors and reviewers are compensated for their work, journals pay authors and reviewers nothing, so more journals and more issues means more profits.

The escalating costs have led to an open-science movement in which journal articles are freely available online, with the publication costs borne by authors (and typically paid by their academic institution or with grant money). There are now several very respectable open-access journals, including PLOS (Public Library of Science).

However, as the open-access model gained traction, unscrupulous publishers moved in to exploit authors trying to win the publish-or-perish battle. Pretty much anyone who has published an article anywhere regularly receives invitations to submit papers for perfunctory review and publication in a predatory journal.

The journal names sound legitimate; indeed, they are often variations on the names of bona fide journals. The *Journal of Economics and Finance* is a legitimate journal; the *International Journal of Economics and Finance* is not. *Journal of Banking and Finance* is legitimate; *Journal of Economics and Banking* is not. *Advances in Mathematics* is legitimate; *International Journal of Advances in Mathematics* is not.

The invitations flatter the recipient, give phony journal impact factors, and promise a fast review process. The last claim is real. There is typically little, if any, review. The only criterion for publication is a willingness to pay the open-access charge.

Invitations from predatory journals pay little attention to one's fields of interests. I have received invitations to write papers on such diverse topics as animal science and veterinary science, osteology and anthropology, properties and prospects of concrete with recycled materials as aggregates or binders. The only thing these topics have in common is that I know nothing about them.

Once you get on a list, you can check-out any time you like, but you can never leave.

The e-mail invitations are often marred by unusual grammar and flawed interpretations of published work. Here are snippets from one that I received.

Dear Colleague!

We have seen your other research manuscripts which are accessible on the internet. We are so interested to invite you to send any of your other new works to our periodical. I would like to let you know that our journal **"Frontiers in Life Science (HFSP)"** is a long lasting BLOCK: N56 peer reviewed periodical printed in Strasbourg, France. Today, as a well BLOCK: N57 scientific periodical, **Frontiers in Life Science** is abstracted and indexed in Science Citation Index (ISI Thomson Reuters) with the five year impact factor of **3.093**. So foremost, we owe our thanks to the authors who have decided to entrust **Frontiers in Life Science** with their best work, allowing us to benefit from the high quality of their scientific BLOCK: N59. We hope to remain worthy of their trust and to BLOCK: N60 providing a rigorous and respected environment for their publications.

Journal Information
Impact Factor: **1.273**
Five-Year-IF: **3.093**
Average Impact Factor: **38.732**

The BLOCKs are formatting instructions that were ignored. The impact factors are fictitious and, if true, would place the journal among the top journals in the world.

Even more worrisome, this is a journal hijacking, which is another arrow in the quiver of nefarious tricks used by predatory journals. There is a legitimate journal titled *Frontiers in Life Science*, but this is not it. More than 100 examples have now been documented of bogus journals creating fake websites that mimic the websites of authentic journals. When a duped researcher "submits" a paper to the imposter journal, it is accepted with minimal fuss and the author is directed to send a publication fee to the huckster.

Two fed-up professors, David Mazières (at NYU at the time, now at Stanford) and Eddie Kohler (at UCLA at the time; now at Harvard) wrote a joke paper titled "Get me off your f**king mailing list" (with the F-word fully spelt out). The paper itself consisted of ten pages of that admonition repeated over and over and over again, along with two graphs incorporating the admonition.

They did not submit this irate paper, but Peter Vamplew, an Australian Professor of Information Technology, responded to an invitation from the predatory *International Journal of Advanced Computer Technology* by submitting the Get Me Off paper, with Mazières and Kohler listed as the authors. Soon after, Vamplew received a congratulatory e-mail informing him that the reviewer had judged the paper "excellent," subject to minor revisions, and that the paper would be published upon receipt of $150 wired to the journal's editor. The reviewer's comments in full:

a. Please Use latest references in order to increase your paper quality.

b. Introduction part is precisely explained.

c. Kindly prepare your camera ready version of paper as per IJACT paper format.

d. Use high resolution images in order to increase the appearance of paper.

Some seemingly reputable publishers have seen the profits to be made from predatory publishing and joined the gold rush. For example, *Expert Systems with Applications*, one of more than 2,600 journals published by Elsevier, now offers authors a $2,640 (plus tax) open-access option. Quantity seems more important than quality. Unlike traditional journals that might publish as few as four issues a year (indeed, some journals have *Quarterly* in their name), *Expert Systems* publishes twenty-four issues a year. In June 2021 I noticed that, unlike traditional journals that might publish ten articles per issue, the November 15, 2021, issue of *Expert Systems* contains eighty-two papers. (Yes, they have so many papers to publish that, in June 2021, they have already published the November 15, 2021, issue.) The November 1, 2021, issue only has thirty-two papers, suggesting that the number of papers published depends on the number of papers submitted. Rejected papers are funneled to "partner publications" (other Elsevier journals) in order to keep the articles and revenue within the Elsevier universe. In 2019 Elsevier made a profit $1.2 billion on revenue of $3.2 billion—yes, a 37 percent margin.

Predatory publishing has become a two-way street. Unscrupulous journals exploit researchers and unscrupulous researchers exploit journals. In 2014 the esteemed journal *Nature* published an article titled "Publishing: The Peer-Review Scam." It recounted the story of Hyung-In Moon, a South Korean medicinal-plant researcher. He had submitted an article to a journal that invites authors to suggest possible reviewers. The intention is to reduce the effort needed by editors to find reviewers for specialized subject matter. Knowing that authors are most likely to suggest personal friends and others who they expect will provide favorable reviews, editors usually add some reviewers who are not on the author's list.

In this case, the editor's suspicions were aroused when it took less than 24 hours for favorable reviews to come in. The editor questioned Moon and he admitted that the reviews took very little time because he had written them himself! It turned out that this was a common practice for Moon. When he suggested reviewers, he used a mix of real and phony names and his own disguised e-mail accounts. A total of twenty-eight papers were subsequently retracted.

One experienced editor, Robert Lindsay, says that he has seen authors recommend not only close friends, but family members and students they are supervising. One brazen researcher recommended herself, using her maiden name. A jaded Lindsay says he that he sometimes uses an author's reviewer suggestions to *rule out* potential reviewers.

Scientific Misconduct

In 2018, Björn Brembs, a professor of neurogenetics at the University of Regensburg, noted the pressure on researchers and the challenge for academic journals:

it has become their task to find the ground-breaking among the too-good-to-be-true data, submitted by desperate scientists, who face unemployment and/or laboratory closure without the next high-profile publication.

One tactic is to p-hack or data-mine one's way to statistical significance. More straightforward is fraud—making up whatever data are needed.

The trickery might be revealed when others try to replicate the results and fail or when others look carefully at the original results and discover anomalies. In the most severe cases, capital punishment for an article is deemed warranted and the author or journal is forced to retract the original paper. The author's reputation may be shredded along with the paper.

Sometimes, one retraction leads to another. When Diederik Stapel, a prominent social psychologist, confessed in 2011 to having made up data, it led to his firing and the eventual retraction of fifty-eight papers. (A fuller discussion is in my book *The AI Delusion*.)

A website named Retraction Watch maintains a searchable list of retracted papers. Figure 13.1 shows their data on the number of retractions each year since 1996, as a percentage of the total number of peer-reviewed science and engineering papers published each year. (Figure 13.1 stops in 2018 because retractions take, on average, about three years.) The steady increase in the retraction rate is distressing.

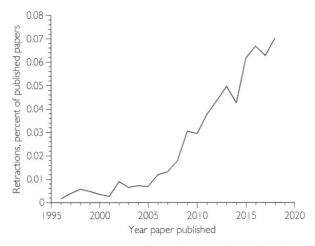

Figure 13.1 Retracted Peer-Reviewed Papers, Percent of Papers Published Each Year.

The number of retractions is small relative to the number of papers published (approaching one in a thousand) but it is surely true that very few cheaters are caught. It is hard work to prove academic misconduct and most researchers rightly reason that advancing their careers through their own work is more important than spending their valuable time destroying someone else's career. It has been estimated that the number of papers published between 1950 and 2004 that should be retracted is between 10,000 (the most optimistic scenario) and 100,000 (a more pessimistic scenario).

The growth in the number of retractions over time that is shown in Figure 13.1 might be explained by the growth in the number of tenth-rate journals. However, several studies have concluded that retractions are more likely for papers published in the most prestigious journals.

Figure 13.2 compares the impact factor (a measure of a journal's reputation) and a retraction index calculated from the fraction of articles that were retracted during the years 2001 through 2010. One explanation for the 0.88 positive correlation is that the payoff for publication in a top-tier journal is irresistibly alluring—it is more tempting to steal a diamond ring than a plastic bracelet. Another possible explanation is that top-tier journals prefer to publish novel and provocative research—which they scrutinize less closely even though it is more likely to be flawed. The infamous Baskervilles paper discussed earlier was surely not vetted carefully, even though the conclusion was surely not correct—that the fourth day of each month is as terrifying to Asian-Americans as being chased down a dark alley by a ferocious dog.

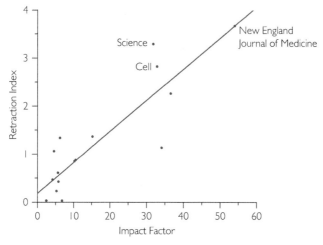

Figure 13.2 Papers published in high-impact journals are more likely to be retracted.

The unreliability of scholarly journals—particularly the most prestigious ones—erodes not only public confidence in science and scientists but also in hiring, promotion, and funding decisions that are based in large part on peer-reviewed publications. It will be calamitous if science comes to be viewed as an elaborate scam intended to enrich the participants.

SCIgen

In the 1920s, Sir Arthur Eddington observed that "If an army of monkeys were strumming on typewriters, they might write all the books in the British Museum." His point was that some things that are not technically impossible might nonetheless be highly improbable. Oddly enough, this example is sometimes used today to make the opposite argument.

In a 1980s debate with a prominent Stanford professor of economics, I argued that Warren Buffett's extraordinary stock market performance was evidence that an investor who processed information better than other investors could do better than the average investor. His response was immediate and dismissive, "Enough monkeys hitting enough keys . . ." His argument was that so many people have been buying and selling stocks over so many decades, that one person is bound to have been so much luckier than the rest as to appear to be a genius—when he is really just a lucky monkey. I was unconvinced. I am not an unbiased observer, but I do believe that some people are better investors than others.

Computers are the perfect instrument for simulating monkeys typing furiously. Many software programs for generating random letters have been written, though approximately 100 percent of the output is gibberish. An easier path to semi-intelligible prose is to generate random *words* instead of random *letters*, and many such programs have also been written.

Tom Mort, a retired environmental field worker, told me about the fun he once had with a random-word generator:

[S]ome of us got involved in setting up a program that followed federal regulations. Our handler was a state agency. Welcome to a new language of Regulese. The jargon and acronyms were astounding in number and intensity.

I suggested that we take a look at [a random text generator] and insert some of this new language into the code.

I had the results typed up and sent on letterhead to our state handlers. The title was "clarification to our plan submission." I figured we'd get a call asking what it was all about. It never came . . . . It was on letterhead and about something real so I imagine it dutifully got stamped and dated as being received and put in the file with all the other stuff we sent them in earnest.

Tom wasn't the first (or last) person to play around with a random text generator. In 2005 three MIT computer science graduate students created a prank program they called SCIgen for using randomly selected words to generate bogus computer-science papers complete with realistic graphs of random numbers. Their goal was to "maximum amusement rather than coherence" and, also, to demonstrate that some academic conferences will accept almost anything.

They submitted a hoax paper titled "Rooter: A Methodology for the Typical Unification of Access Points and Redundancy," to the World Multiconference on Systemics, Cybernetics and Informatics that would be held in Orlando, Florida. The gibberish abstract read:

Many physicists would agree that, had it not been for congestion control, the evaluation of web browsers might never have occurred. In fact, few hackers worldwide would disagree with the essential unification of voice-over-IP and public-private key pair. In order to solve this riddle, we confirm that SMPs can be made stochastic, cacheable, and interposable.

Among the bogus references was,

Aguayo, D., Aguayo, D., Krohn, M., Stribling, J., Corbato, F., Harris, U., Schroedinger, E., Aguayo, D., Wilkinson, J., Yao, A., Patterson, D., Welsh, M., Hawking, S., and Schroedinger, E. A case for 802.11b. *Journal of Automated Reasoning*, 904 (Sept. 2003), 89–106.

D. Aguayo appears three times in the list of authors and E. Schroedinger appears twice. Aguayo is one of the MIT pranksters and Schroedinger is an apparent reference to Erwin Schroedinger, the Nobel Laureate physicist who died in 1961, forty-two years before the referenced paper was said to have been written.

Nonetheless, the conference organizers quickly accepted the fake paper. The students then revealed their prank and it was covered by CNN, BBC, and other prominent media, which forced the conference to withdraw its acceptance.

Undaunted, the students rented a conference room in the hotel where the conference was being held and held sessions in which they presented randomly generated slides that they had not seen before their presentation and made a video documenting their continuing prank.

They have now gone on to bigger and better things, but SciGen lives on. Believe it or don't, some resourceful (desperate?) researchers bolster their CVs by using SCIgen to create papers that they submit to conferences and journals.

Cyril Labbé, an energetic and enterprising computer scientist at the University of Grenoble, wrote a program to detect hoax papers published in real journals. Working with Guillaume Cabanac at the University of Toulouse, he found 243 bogus published papers written entirely or in part by SCIgen. A total of nineteen publishers were involved, all reputable and all claiming that they only publish papers that pass rigorous peer review. One of the embarrassed publishers, Springer, subsequently announced that it was teaming with Labbé to develop a tool, SciDirect, that would identify nonsense papers. The obvious question is why such a tool is needed. Is the peer-review system so broken that reviewers cannot recognize nonsense when they read it?

Even more ominously, GPT-3 and other LLMs are now capable of generating papers sufficiently coherent and well-written that they can be used for school essay assignments and academic journal articles. Some have lamented (and others have celebrated) that learning to write, like learning to do long division, may now be obsolete. One not-so-slight problem is that these programs do not, and do not try to, understand what words mean. As with chatbots, they are nothing more than fluent spouters of bullshit. It is unfortunate that so much effort has been spent creating systems designed to deceive.

Gaming Citation Counts

In 2010 Labbé showed how citation counts could be inflated. In a few short months, he elevated an imaginary computer scientist (Ike Antkare, pronounced, "I can't care") to "one of the greatest stars in the scientific firmament." Labbé used SCIgen to create a fake paper, purportedly authored by Antkare, which referenced real papers. Then Labbé used SCIgen to generate 100 additional bogus papers supposedly authored by Antkare, each of which cited itself, the other ninety-nine papers, and the initial fake paper. Finally, Labbé created a web page listing the titles and abstracts of all 101 papers, with links to pdf files, and waited for Google's web crawler to find the bogus cross-referenced papers.

The Googlebot did its job and Antkare was credited with 101 papers that had been cited by 101 papers, which propelled him to twenty-first on Google's list of the most cited scientists of all time, behind Freud but well ahead of Einstein, and first among computer scientists.

Even if researchers do not do an all-out Ike Antkare, they can still easily game citation metrics. In every paper they write, they can cite as many of their own papers as the editors will let them get away with. Journals can also game citation counts by publishing lots of papers that cite papers previously published in the journal. On more than one occasion, I have had journal editors ask me to add references to articles published in their journal.

Out of curiosity I randomly selected a paper from the *Expert Systems* journal mentioned earlier and found that it had nine references to other *Expert Systems* papers. If a journal publishes 1,200 papers a year, each with nine citations to other papers published in the journal, authors will average nine citations per paper and the journal will average 10,800 citations per year, even if their articles are never cited by papers published in any other journal. Publishers like Elsevier that have a large portfolio of journals can also boost citation counts by encouraging authors to cross-cite other journals in the portfolio. *Expert Systems* is, in fact, among the top-five artificial-intelligence journals in terms of total citations, but ranks twenty-seventh when the citations are weighted by the importance of the journals that the citations come from.

What to do? What to do? Publication counts and citation indexes are too noisy and too easily manipulated to be reliable. Nor can we evaluate research simply by noting the journals that publish the results. There is so much noise in the review process that lots of great papers have been published in lesser journals and many terrible papers have been published in the most respected journals.

John P. A. Ioannidis's paper "Why Most Published Research Findings Are False" has been cited nearly 10,000 times despite being published in *PLOS Medicine*, a good-but-not-great open-access journal. On the other hand, Chapter 7 discussed several flawed papers published in the *British Medical Journal*, a truly great journal.

The best solution is to have experts actually read the research done by applicants for a job, promotion, or grant. Simple counts and indexes won't do. An old aphorism in academia is that "Deans can't read, but they can count." A better one is "Not everything that can be counted counts, and not everything that counts can be counted."

The Irony

It is natural to assume that peer review ensures that published research papers are of sufficiently high quality to have been certified by experts in the field. Unfortunately, peer review is no panacea. Reviewers understandably assume that their colleagues are honest and that the reported calculations are an accurate reflection of the research that was done. Expert reviewers, in particular, are loath to spend the time needed to scrutinize the data and check the calculations even if the data are available, which they often aren't.

Science created the peer-review model to weed out unreliable research. When it is revealed that some published papers are based on tweaked or invented data, that some papers report incorrect calculations, that some papers are written by computers, that publishing can be gamed, it diminishes our faith in scientific research—and science.

CHAPTER 14

The Replication Crisis

obel Laureate James Tobin wryly observed that the bad old days, when researchers had to do calculations by hand, were actually a blessing. The calculations were so hard that people thought hard before calculating. They put theory before data. Today, with terabytes of data and lightning-fast computers, it is too easy to calculate first, think later.

Computer algorithms are terrible at identifying logical theories and selecting appropriate data for testing these theories, but they are really, really good at data mining—rummaging through data for statistically significant relationships. They are also really good at aiding and abetting p-hacking—torturing data to identify a pattern with a low *p*-value.

The overriding problem with p-hacking and data mining is that the reported results so often vanish when tested with fresh data—a disappearing act that contributes to the replication crisis that is undermining the credibility of scientific research.

Even if researchers don't p-hack individually, there can be collective p-hacking. Suppose 100 researchers each test a different worthless theory. We expect five, by luck alone, to have *p*-values below 0.05. The other ninety-five tests disappear without a trace—the same as if one researcher tested 100 worthless theories and only reported the five best results. This is called *publication bias*, in that the statistically significant results are the ones that get published, while the insignificant results are put in a file drawer and forgotten.

A 2013 study found that half of all clinical medical trials are never published. Sometimes, the results contradict the researchers' expectations. More often, the results are not statistically significant and, therefore, considered uninteresting, even though a full assessment of the efficacy of a medical treatment should consider all the evidence.

A research team once conducted nine studies on the use of a nasal spray to attract the opposite sex. Four studies showed a positive effect; the other five studies did not. They submitted all nine articles to medical journals. The four articles with positive effects were accepted while four of the five articles with statistically insignificant effects were rejected multiple times and were never published. In a subsequent article subtitled, "Opening the File Drawer of One Laboratory," the team wrote that "our publication portfolio has become less and less representative of our actual findings." and recommended that researchers "get these unpublished studies out of our drawers and encourage other laboratories to do the same."

Bitcoin Redux

In this book's Introduction, I discussed a data-mining study that estimated 810 correlations between bitcoin returns and seemingly randomly chosen economic variables and found sixty-three relationships with p-values below 0.10 (and thus were labeled "statistically significant at the 10 percent level"). This is somewhat fewer than the eighty-one statistically significant relationships that would be expected if the authors had just correlated bitcoin returns with random numbers. They went on a massive fishing expedition and came home empty handed.

Owen Rosebeck and I redid their analysis to see how many of these sixty-three relationships held up out of sample, during the 14 months after they completed their study. Not surprisingly, seven (approximately 10 percent) did—again what we expect with purely random numbers.

We submitted our paper to a well-regarded journal and were given an interesting rejection. One reviewer said that we were "incorrect" when we tested only the sixty-three relationships that the authors had found to be statistically significant. We should also have used the out-of-sample data to test the 747 relationships that were not statistically significant in-sample!

What if you found a significant result where they found nothing? That is worth knowing. This mistake is repeated several times throughout the paper.

What a wonderful example of how researchers can be blinded by the lure of statistical significance. Here, in a study of the relationship between bitcoin returns and hundreds of unrelated economic variables, this reviewer, respected enough to be used by a good journal, believes that it is worth knowing if statistically significant relationships can be found in some part of the data even if we already know that none of these relationships replicate in the rest of the data.

That is the pure essence of the replication crisis—the ill-founded belief that a researcher's goal is to find statistically significant relationships, even if the researcher knows in advance that the relationships do not make sense and do not replicate.

A crucial step for tamping down the replication crisis is for researchers to recognize that statistical significance is not the goal. Real science is about real relationships that endure and can be used to make useful predictions with fresh data.

Get Your 8 (or 5) Hours of Sleep

Mathew Walker is a professor of neuroscience and psychology and founder of the Center for Human Sleep Science at the University of California, Berkeley. He has become famous for his books and a TED talk promoting the importance of sleep for health and performance. He even got a job at Google as a "sleep scientist."

Walker has a receptive audience because his arguments make sense and he is entertaining. In one of his books, Walker used a graph similar to Figure 14.1 to show that a study done by others had found that adolescent athletes who sleep more are less likely to be injured.

The figure is compelling, but there are several potential problems. The sleep data were based on 112 responses to an online survey of athletes at a combined middle school/high school. The injury data were from logs of students who came to the school athletic trainer's room for "evaluation and/or treatment." Overall, 64 of the 112 athletes made a total of 205 visits.

Online surveys are notoriously suspect and recollections of the average hours of sleep are likely to be unreliable. The training room data spanned a 21-month period but, nonetheless, seem high. Perhaps some middle school/high school students preferred the trainer's room to being at practice.

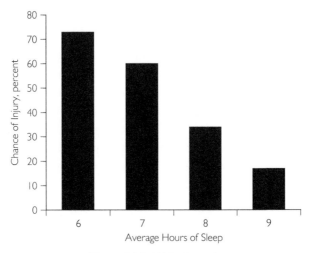

Figure 14.1 Walker's graph.

There is also a question of which way causation goes. If, in fact, people who sleep less have more injuries, is it because lack of sleep makes them more injury-prone or because injuries make it harder to sleep? Or perhaps high-energy people need less sleep and also play sports more recklessly or play more reckless sports?

The most serious problem, however, was not that Walker was relying on flimsy data. Finnish fitness blogger Olli Haataja took the trouble to read the original study and discovered that the published graph (Figure 14.2) showed five sleep categories, instead of the four Walker reported. Walker had omitted the 5-hour category, which contradicted his argument! It is difficult to explain this omission as anything other than a deliberate attempt to misrepresent the results of the study.

Independent researcher Alexey Guzey put this falsification on a website he maintains, listing errors in Walker's work. Andrew Gelman then wrote about Walker's distortion in a blog and in an article co-authored with Guzey, arguing that "When it comes to unethical statistical moves, hiding data is about as bad as it gets."

A software developer named Yngve Hoiseth was so incensed that he contacted the University of California to report that Walker had been guilty of research misconduct, as defined by the university:

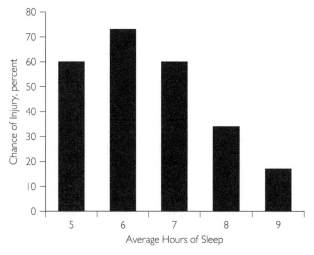

Figure 14.2 The original graph.

Falsification is manipulating research materials, equipment, or processes, or changing or omitting data or results such that the research is not accurately represented in the research record.

Walker clearly misrepresented the research record by omitting some results. The university staunchly defended its celebrity professor:

In conversation with Walker and with the professor who conducted the inquiry, the conclusion was that the bar omitted from the graph on the book did not alter the findings in an appreciable way, and more importantly, that the bar was not omitted in order to alter the research results.

Gelman subsequently asked the obvious question:

If the removal of the bar from the graph didn't matter, then why did you remove the damn bar?

This misrepresentation of results does not seem to be an isolated incident. Markus Loecher, a professor of mathematics and statistics at the Berlin School of Economics and Law, reported similar mischief in Walker's TED talk, "Sleep is your superpower." At one point, Walker made this dramatic argument:

I could tell you about sleep loss and your cardiovascular system, and that all it takes is one hour. Because there is a global experiment performed on 1.6 billion people across

70 countries twice a year, and it's called daylight saving time. Now, in the spring, when we lose one hour of sleep, we see a subsequent 24-percent increase in heart attacks that following day. In the autumn, when we gain an hour of sleep, we see a 21-percent reduction in heart attacks. Isn't that incredible? And you see exactly the same profile for car crashes, road traffic accidents, even suicide rates.

After listening to this, Loecher wrote:

Now I tend to be sensitive to gross exaggerations disguised as "scientific findings" and upon hearing of such a ridiculously large effect of a one-day-one-hour sleep disturbance, all of my alarm bells went up!

He contacted Walker and was told that the source of these claims was a study of Michigan heart attacks following four spring and three fall daylight savings time changes, not "a global experiment performed on 1.6 billion people across 70 countries."

Table 14.1 shows that there was indeed a 24 percent increase after the spring change and a 21 percent decrease after fall change, as Walker stated, but these fluctuations did not happen, as Walker claimed, the "following day." Saturday night is when we have one less or one more hour of sleep and the following day is Sunday. The spring increase in the Michigan data was on a Monday and the fall decrease was on a Tuesday, two seemingly random days—except for the fact that they were the only days to have had p-values (0.011 an 0.044, respectively) below 0.05. The Sunday after the time changes, the number of heart attacks went in the opposite direction.

Perhaps the sleep effects are felt during the entire week following the time changes. Nope. The Michigan authors clearly state that "There was no difference in the total weekly number of [heart attacks] for either the fall or spring time changes." The authors also caution that they had used multiple statistical procedures and that

No adjustments were made for multiple comparisons, and the analysis is intended to be exploratory in nature. As such, nominally significant results should be interpreted as hypothesis generating, rather than confirmatory evidence.

Walker took the inconclusive results of a small exploratory study and inflated it into a claim that there were well-documented worldwide surges and declines in heart attacks on the Sunday following time changes.

Loecher also noted that, unsurprisingly, the Michigan results did not hold up in other data. He contacted one of the authors of the Michigan study and

Table 14.1 *Relative Risk of Heart Attack during the Week after Daylight Saving Time Changes*

	Spring time changes	Fall time changes
Sunday	0.97	1.02
Monday	1.24	0.94
Tuesday	0.98	0.79
Wednesday	0.97	0.94
Thursday	0.97	1.10
Friday	0.97	0.91
Saturday	1.04	1.15

was told that the authors "looked for the same signal in more recent data and it is markedly attenuated and no longer significant." Loecher also reported—and this, too, should come as no surprise—that

BTW, I was unable to find any backing of the statement on "exactly the same profile for car crashes, road traffic accidents, even suicide rates" in the literature.

Boring results do not receive worldwide publicity, TED-talk invitations, or Google jobs—which is why it is tempting to misrepresent and exaggerate in order to impress and entertain. The social cost is that misrepresentations and exaggerations undermine the credibility of scientists by making them seem more like snake-oil peddlers than dispassionate scientists.

Fake It 'til You Make It—the Power Pose Parable

A 2010 paper published as the lead article in a top-tier psychology journal advised that "a person can, by assuming two simple 1-min poses, embody power and instantly become more powerful." The researchers had forty-two people assume two positions for one minute each—either high-power (Figure 14.3) or low-power poses (Figure 14.4).

Saliva samples were used to measure the dominance hormone testosterone and the stress hormone cortisol. Risk taking was gauged by a willingness to take a bet with a 50 percent chance of winning $2 and a 50 percent chance of losing $2. Feelings of power were measured by stating on a scale of 1 to 4 how "powerful" and "in charge" they felt.

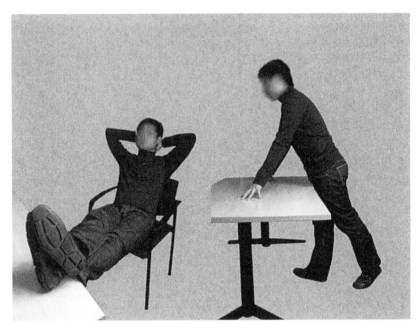

Figure 14.3 High-power poses.
Credit: Sage Publishing.

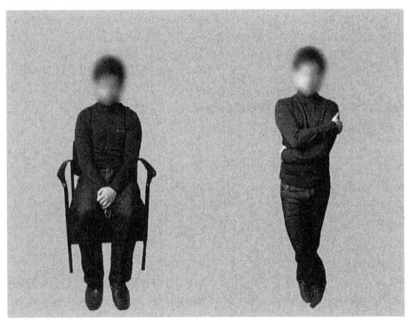

Figure 14.4 Low-power poses.
Credit: Sage Publishing.

Compared to the low-power poses, those who took high-power poses had an increase in testosterone, a reduction in cortisol, an increased willingness to take risks, and increased feelings of power. All the p-values were less than 0.05 and the authors concluded that

a simple 2-min power-pose manipulation was enough to significantly alter the physiological, mental, and feelings states of our participants. The implications of these results for everyday life are substantial.

The implications were certainly substantial for one of the co-authors, Amy Cuddy. She wrote a best-selling book, gave one of the most-watched TED talks of all time, and became a celebrity speaker with fees of $50,000 to $100,000.

In her TED talk, Cuddy is confident, emotional, and charismatic while she tells the audience that her advice is backed by solid science. One of her memorable lines is don't just fake it till you make it, but fake it till you become it. A giant screen projects a picture of Wonder Woman standing with her legs spread and hands on her hips while Cuddy tells her audience:

Before you go into the next evaluative situation, for two minutes, try doing this in an elevator, in a bathroom stall, at your desk behind closed doors . . . . Get your testosterone up. Get your cortisol down.

It turned out that it is not so easy to get your testosterone up and your cortisol down. In 2015 there was a published report by a group of researchers who wanted to explore gender differences in the benefits from power poses. However, using a much larger sample (200 subjects), they found only a small increase in self-reported feelings of power and no effects on hormone levels or behavior.

In 2017 the journal *Comprehensive Results in Social Psychology* (CRSP) published the results of several large studies, each using a peer-reviewed pre-registration that specified, in advance, how the studies would be conducted in order to limit the possibilities for p-hacking. These studies did not find evidence of hormonal or behavioral changes, but did find modest effects on feelings of power.

Dana Carney was the lead author of the original power-posing paper and she provided detailed feedback on the pre-registration plans for the CRSP papers. In September 2016, she posted a statement on her faculty website at the University of California, Berkeley:

As evidence has come in over these past 2+ years, my views have updated to reflect the evidence. As such, I do not believe that "power pose" effects are real.

Carney also revealed that there had been some p-hacking in the original study. Here are a few excerpts from her statement:

> The sample size is tiny.
>
> The data are flimsy. The effects are small and barely there in many cases.
>
> We ran subjects in chunks and checked the effect along the way. It was something like 25 subjects run, then 10, then 7, then 5. Back then this did not seem like p-hacking. It seemed like saving money.
>
> Some subjects were excluded on bases such as "didn't follow directions." The total number of exclusions was 5.
>
> Subjects with outliers were held out of the hormone analyses but not all analyses.
>
> Many different power questions were asked and those chosen were the ones that "worked."

The p-hacks may well explain why the initial results were statistically significant but did not replicate.

In Chapter 5, we saw that adding observations or deleting outliers can substantially increase the probability of false positives. The power-pose study did both. Here, the original study started with twenty-five subjects, then added ten more, then seven, then five. To calculate the true chances of a false positive, we need to make an assumption about what the researchers would have done if forty-seven subjects had not been sufficient. Figure 14.5 shows the false-positive probabilities if they had continued adding batches of five subjects for maximum sample sizes up to 102.

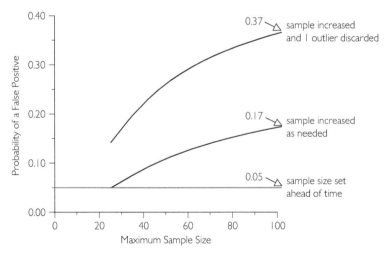

Figure 14.5 Statistical significance at the cost of false positives.

Figure 14.5 also shows how the false-positive probability is affected by discarding an outlier in order to reduce the p-value. With a 102-subject stopping point, the flexible sample size pushes the probability of a false positive from 5 to 17 percent. Discarding one outlier as needed further pushes the false-positive probability to 37 percent. A willingness to discard more than one outlier would increase the probability of a false-positive even more.

I don't know the extent to which the discarded observations in the original power-pose study reduced the p-values, but it is troubling that there were forty-seven subjects, not the forty-two reported, with five people excluded for vague reasons like "didn't follow directions" and that another observation was omitted from the hormone results, but not the other results.

In addition, the reporting of only a selected subset of the power questions was almost surely a substantial p-hack. Altogether, the false-positive probability was most likely well over 50 percent.

Carney's willingness to acknowledge the p-hacks and to support efforts to redo the power-pose tests is convincing evidence that the p-hacks were well-intentioned. This was how a lot of research was done at the time. Joseph Simmons, Leif Nelson, and Uri Simonsohn, who have persuasively documented the dangers of p-hacking, have written that

We knew many researchers—including ourselves—who readily admitted to dropping dependent variables, conditions, or participants so as to achieve significance. Everyone knew it was wrong, but they thought it was wrong the way it's wrong to jaywalk. We [now know that it] was wrong the way it's wrong to rob a bank.

A 2011 survey of 2,000 research psychologists found that 72 percent admitted to having used statistical significance to decide either to collect additional data or to stop collecting data; 38 percent to "deciding whether to exclude data after looking at the impact of doing so on the results;" 46 percent to selectively reporting studies that "worked," and 27 percent to "reporting an unexpected finding as having been predicted from the start."

Criticism of Cuddy is unfair to the extent that she was doing what others did at the time—indeed what her professors taught her to do. On the other hand, she was quick to exaggerate fragile results based on a tiny sample and slow to acknowledge that larger studies did not support her conclusions. Audiences want simple, powerful messages, and that is what she gave them. She seemed to be following her own advice when she confidently promoted her power-pose story: fake-it-till-you-make-it, fake-it-till-you-become-it.

What's left of power posing? Artificial poses don't seem to affect people's hormones or their behavior. Even the possibility of a modest effect on feelings is suspect because the subjects who know what researchers expect may change their behavior to meet those expectations. Here, people who are instructed to assume very unusual and possibly awkward poses and then asked how powerful they feel may well know the desired answer. This explanation is supported by one study that found that people who had viewed Cuddy's TED talk were more likely to report feeling powerful after assuming a high-power pose.

Another issue is that the power-pose study compared high-power posing to low-power posing, but the TED talk advises high-power posing instead of a normal body position. Perhaps the differences in power feelings found in the study are due to the negative effects of crossing one's arms and legs rather than the positive effects of spreading one's arms and legs. This is a well-known "poison-medicine" problem—if sick people take a medicine and live while others drink poison and die, this doesn't mean that the medicine worked.

The real value of the power-posing parable is that it is a compelling example of how p-hacking has fueled the replication crisis that has shaken science. The original study was unreliable because the goal was to deliver a simple media-friendly message and the consequences of p-hacking were not yet widely appreciated. An enormous amount of valuable time and resources were then spent showing that the study was unreliable. What we have learned from this parable is that it is far better to do studies correctly the first time—and that is what serious researchers are striving to do now.

The Times They are A-Changin'

In her TED talk, Cuddy called Susan Fiske, her dissertation adviser and past president of the Association for Psychological Science, her "angel adviser." Fiske returned the favor in 2016 with accusations of "methodological terrorism" and a rousing defense of researchers caught up in the replication crisis:

[T]he new media (e.g., blogs, Twitter, Facebook) can encourage a certain amount of uncurated, unfiltered denigration. In the most extreme examples, individuals are finding their research programs, their careers, and their personal integrity under attack . . . . I have heard from graduate students opting out of academia, assistant professors afraid to come up for tenure, midcareer people wondering how to protect their labs, and senior faculty retiring early, all reportedly because of an atmosphere of methodological intimidation.

Several online responses questioned her conclusions:

> Like Fiske, I know many young scientists leaving research . . . . [P]ublic shaming or harassment is not a factor in any case I know of (unless you count people who were involved in misconduct). It's simply not an issue on their radar at all.
>
> I know of exactly 0 young scientists who have left research because of public criticism of established research/ers. However, I know of at least a handful of people in my cohort or in adjacent cohorts who have left research because:
>
> 1. They were disillusioned by the amount of "bad science" . . .
>
> 2. Because they felt like the cards were stacked against them and that staying in science was against their own best interest—the low-hanging fruit have been picked (but perhaps via flawed studies that no one cares about validating)
>
> To be clear, no one in my cohort cares about the careers of established older PIs. Many of us have little faith in the quality of their work, in the strength of their commitment to science when there is potential for damaging their careers and/or "professional reputation" or challenging their "expertise," and would be happy to see them and the system that supports them collapse so that we could do science we have faith in.

Yes, some members of an earlier generation are understandably upset about criticisms of the methods they used to advance their careers. Andrew Gelman has written that Fiske is "seeing her professional world collapsing. It's scary, and it's gotta be a lot easier for her to blame some unnamed 'terrorists' than to confront the gaps in her own understanding of research methods."

Who Put My Keys in the Refrigerator?

Fiske was journal editor for some now-infamous papers (including the him-micane paper I will discuss later in this chapter) and co-authored at least one dodgy paper, which concluded that the elderly are viewed more warmly if they are perceived to be incompetent. Eighteen of the participants (the control group) read a brief story about an elderly man named George. For an additional eighteen participants (the high-incompetence group), the story added a sentence telling how George has been having memory problems and recently couldn't remember where he put his keys. For the final nineteen participants (the low-incompetence group), the added sentence described George as priding himself on his perfect memory. After they read the story,

> Participants then indicated how much George possessed each of a series of personality traits, on a scale from 1 (not at all) to 9 (very). We included one manipulation check (competence) and three items designed to assess warmth (warm, friendly, and good-natured).

The average ratings were 7.47 for the high-incompetence elderly, 6.85 for the low-incompetence elderly, and 6.59 for the control group. The low-incompetence group was actually rated warmer than the control group, contrary to the authors' stated expectations.

It seems odd that there were eighteen, eighteen, and nineteen subjects in the three groups rather than, say, twenty in each. Perhaps, as in the power-pose paper, some subjects were added and some data were discarded. Also, the "series of personality traits" might have included some questions that were discarded for not being statistically significant. There is no way to know unless an author admits to such mischief but my suspicions were aroused here by the fact that Cuddy was a co-author of both the power-pose paper and the incompetent-elderly paper.

I queried Fiske and received a lengthy response, the key elements being: "I did not run the study . . . I assume but can't check that what the paper says is what we did." I e-mailed Cuddy and the third co-author and received no response.

The authors did an overall statistical test and found the differences among the three groups were barely significant ($p < 0.03$) but that the pairwise differences between the high/low and high/control groups were highly significant, $p = 0.00001$ and 0.00000000000004, respectively. Most introductory statistics students would realize that something is wrong here. If the p-values for the pair-wise tests are extremely small, the overall p-value should be small, too. The authors and journal reviewers should have noticed this obvious contradiction.

Nick Brown reverse-engineered the results and found that if the overall p-value is correct, then the correct pairwise p-values for the high/low and high/control groups are 0.0811 and 0.0024.

To their credit, the authors published a correction, admitting this error:

The Results on p. 275 should read (changes in bold):
"Paired comparisons supported these findings, that the high-incompetence elderly person was rated as warmer than both the low-incompetence **elderly target (marginally)** and control **elderly target (significantly)**, $t(35) = 1.79$, $p = .08$, and $t(34) = 3.29$, $p<.01$, respectively."
This correction does not change the conclusions of the article.

Whether or not this error changes the conclusions is a matter of opinion, not fact. The article's research hypothesis is that "We might expect a competent elderly person to be seen as less warm than a reassuringly incompetent elderly person." By their p < 0.05 criterion, they did not find statistically significant evidence that the incompetent elderly are perceived more warmly (p = 0.08).

It is also noteworthy that the authors did not see any difference between p-values of 0.00000000000004 and 0.0024. The mindset seems to be that as long as the p-value is below 0.05, it doesn't matter what it is. Gelman commented that "When the authors protest that none of the errors really matter, it makes you realize that, in these projects, the data hardly matter at all."

Many researchers now recognize the consequences of p-hacking and data mining and are working hard to restore the credibility of scientific research. Gelman has described the old way as the "find-statistical-significance-any-way-you-can-and-declare-victory paradigm," and written that "I can see that to people such as Fiske who'd adapted to the earlier lay of the land, these changes can feel catastrophic."

Eli Finkel, a social psychologist at Northwestern University, wrote that

It was like we had been having a big party—what big, new, fun, cool stuff can we discover? And we forgot to double-check ourselves. And then the reformers were annoyed, because they felt like they had to come in after the fact and clean up after us. And it was true.

Michael Inzlicht, professor of psychology at the University of Toronto, spoke for many when he wrote that

I want a better tomorrow, I want social psychology to change. But, the only way we can really change is if we reckon with our past, coming clean that we erred; and erred badly . . . . Our problems are not small and they will not be remedied by small fixes. Our problems are systemic and they are at the core of how we conduct our science.

Replication Studies

Brian Nosek's Reproducibility Project attempted to replicate 100 studies that had been published in what are arguably the top three psychology journals. Only thirty-six continued to have p-values below 0.05 and to have effects in the same direction as in the original studies. Even among the few that could be replicated, the effects were generally smaller in the follow-up study than in the original reported results.

In December 2021, the Center for Open Science (co-founded by Nosek) and Science Exchange reported the results of an 8-year project attempting to replicate twenty-three highly cited in-vitro or animal-based preclinical cancer biology studies. The twenty-three papers involved a total of fifty experiments and 158 estimated effects. Only 46 percent replicated and the median effect size was 85 percent smaller than originally estimated. (That is a major decline effect!)

Colin Camerer led a team that looked at eighteen experimental economics papers published in two top economics journals. Only eleven (61 percent) were successfully replicated. Camerer also led a team that attempted to replicate twenty-one experimental social science studies published in two leading general science journals (*Nature* and *Science*) and found that only thirteen (62 percent) continued to be statistically significant and in the same direction with fresh data.

If it is true that the top journals have the most rigorous review process and publish the most reliable research (a debatable assumption!), then the replication rates are likely to be even lower for research published in lesser journals. Ioannidis guesstimates that 90 percent of published medical research is flawed in that treatments reported to be beneficial are less beneficial than reported and sometimes worthless or worse.

Anecdotal evidence comes from a survey of 1,500 scientists conducted by *Nature*. More than 70 percent reported that they had tried and failed to replicate another scientist's experiment and more than half had tried and failed to replicate some of their own studies!

Replication Markets

An interesting side study was done while Nosek's Reproducibility Project was under way. Approximately two months before forty-four of the replication studies were scheduled to be completed, forty-four auction markets were set up for researchers in the field of psychology to bet on whether the replication would be successful. The markets were open for two weeks and the people doing the studies were not allowed to participate. A replication was considered to be successful if the main result was found to have a p-value less than 0.05 and to be in the same direction as the original result (or to have a p-value above 0.05 for the one paper that initially reported a p-value larger than 0.05).

The final market prices indicated that people believed that there was, on average, only a 55 percent chance of a successful replication. Of the forty-one studies that were completed on time, nineteen (46 percent) were given less than a 50 percent chance of replicating. Even that dismal expectation turned out to be too optimistic as twenty-five (61 percent) did not replicate.

For me, the most important takeaway from the replication markets and survey probabilities is how skeptical psychology researchers are of psychology research.

In 2015 I had been invited to Sci Foo, an annual gathering of around 250 scientists, writers, and policy makers at the Googleplex, Google's corporate headquarters in Mountain View, California. One hot topic was the replication crisis that was creating havoc in so many fields. A prominent social psychologist said that his field is the poster child for the replicability crisis and added that "My default assumption is that anything published in my field is wrong." That's just one opinion and it may have been deliberately hyperbolic, but these data support his cynicism.

The more general point, and it goes far beyond social psychology, is that if empirical results are so fragile that even researchers do not take them seriously, what are lay people to think?

Zombie Studies

A 2021 study found that the non-replicated papers in the three replication projects described in the section "Replication Markets" were cited by other researchers more often than were the papers that were successfully replicated! Table 14.2 shows that this disparity was largest for papers published in the highly prestigious general science journals *Nature* and *Science*. This study also found that the citation gap tended to grow over time and persisted even after the replication projects had identified the papers as flawed.

A plausible explanation is that entertaining and memorable results are often surprising, and there is a good reason why they are surprising—they are incorrect. The allure of conclusions that are novel/interesting/provocative leads some researchers to stretch to find statistical significance, leads journals to publish flawed papers, and leads other researchers to cite these faulty papers more often.

Table 14.2 *Average Number of Citations*

	Not-replicated papers	Replicated papers
Nature/Science	638 (68%)	301 (32%)
Economics	239 (63%)	141 (37%)
Psychology	187 (52%)	175 (48%)

One example that I see every hurricane season is social media reliably abuzz with reports of a debunked study that claimed that hurricanes given feminine-sounding names are deadlier than those given masculine-sounding names. (Full disclosure: I was one of the debunkers.)

The authors of what is now called the *himmicane* paper were not arguing that female hurricanes are inherently deadlier but that sexist humans die because they don't take female hurricanes seriously. There were a lot of problems with the study, starting with the fact that the results hinge on the inclusion of 1953–1978 data, when all hurricanes had female names and hurricanes were, on average, stronger; buildings were weaker; and there was less advance warning.

In addition, there were a large number of forking paths in the original research with the authors seemingly choosing a route that led them to their desired conclusion. Speaking plainly, this study was a mass of hot air based on a questionable analysis of inappropriate data and it didn't hold up with fresh data. A fuller discussion of the hurricane/himmicane debacle is in my book *The AI Delusion*.

My point here is simply that flawed studies that have provocative conclusions may live on in popular culture long after their flaws are exposed. They become zombie studies.

Another, even more disastrous, example is the vaccine scare created by the British doctor Andrew Wakefield. His 1998 co-authored paper in the prestigious medical journal *The Lancet* claimed that twelve normal children had become autistic after being given the measles, mumps, and rubella (MMR) vaccine. Wakefield held a press conference before the paper was published, announcing his findings and calling for the suspension of the MMR vaccine.

Soon, problems were uncovered with Wakefield's study—problems so serious that *The Lancet* retracted the article in 2010, with an editorial comment, "it was utterly clear, without any ambiguity at all, that the statements in the paper were utterly false." The *British Medical Journal* called the Wakefield study "an elaborate fraud," and the UK General Medical Council barred Wakefield from practicing medicine in the UK.

Unfortunately, the damage was done. Far more people know about Wakefield's claims than know about the retraction and censure. Millions of parents have refused to allow their children to be given the MMR vaccine, and thousands of unvaccinated children have died from measles, mumps, and rubella because of this zombie study.

The Irony

Our lives have been enriched immensely by the scientific method's insistence that beliefs and theories should not be accepted uncritically, but tested empirically. Unfortunately, clever researchers have found ways to game the system by data mining and p-hacking in order to achieve statistical significance. The consequence is the replication crisis that threatens to undermine the reputation of science and scientists.

Real science still moves forward with honest tests of plausible theories that can be replicated with additional tests. Our challenge—and the challenge for scientists—is to distinguish real science from junk science.

CHAPTER 15

Restoring the Luster of Science

S cience is a powerful engine that has allowed us to live longer and better lives. We travel near or far in cars that are heated or cooled as needed while we are entertained by music or podcasts. We stay awake well past dark in our well-lit homes, reading books, watching television, playing computer games, and communicating with friends and family via our smartphones and the Internet. We have comfortable furniture, indoor plumbing, and appliances that clean our dishes, clothes, and homes. We eat food that we don't have to grow ourselves. We keep food fresh in refrigerators and cook food with convenient stoves and ovens—or we go to restaurants or have food delivered to our homes. When we are sick or injured, we can receive medical treatments that are safe and effective.

Unfortunately, too many people no longer trust science or scientists. There are three fundamental reasons for the diminished respect for science:

Disinformation

Data torturing

Data mining

In each case, there are ways to restore the luster of science. It won't be easy, but it is worth doing.

Disinformation

A distrust of elites in general and scientists in particular is spread and magnified by the Internet swamp of fake stories, misleading videos, and manipulated social media. For too many years, I have hoped that social media was a passing fad that would wither away as people come to recognize the thousands of hours they have wasted on gossip and lies and resent being manipulated. The world seems to be moving farther from that optimistic hope.

Filter Bubbles

Internet companies amass information about our clicks, search history, demographics, and more while they engage us by recommending products, stories, and links. This makes sense. Internet companies want users to spend more time on their platforms, following link after link, seeing advertisement after advertisement.

The algorithms value engagement—stories people will read, videos they will watch, tweets they will share. Provocative links are promoted—and things that are provocative are often exaggerated, misleading, or downright lies. In addition to promoting links that titillate, algorithms select content that confirms biases and fuels hyper-partisanship by luring people into "filter bubbles" in which they mostly read and watch things that support their worldview. Users seldom leave their bubbles—indeed, they don't know they are inside a bubble—because the algorithms keep feeding them more of the same. If users are aware of people with other viewpoints, the others are deemed enemies—uninformed and clueless.

It is difficult for society to fight back without endangering basic rights. There is surely potential value in removing false and malicious content from the Internet, but who is to decide what is true and what is false, what is benign and what is malicious?

Totalitarian governments trample freedoms by censoring media, but democracies should be wary. Freedom of speech in general and freedom of the press in particular can reveal government secrets and abuses—so that governments can be held accountable and, knowing this, will be more likely to serve citizens instead of exploiting them.

Freedom of Speech

A cornerstone of democracy is the right of the governed to speak freely without fear of being jailed or otherwise punished for criticizing the government. As expressed in the First Amendment to the U.S. Constitution,

Congress shall make no law . . . abridging the freedom of speech, or of the press.

Freedom of speech is, however, not absolute. The First Amendment does not protect speech directly related to criminal activity, nor does it protect false advertising, fraud, harassment, threats of violence, libel, or even unwanted robocalls. Many cases are ambiguous and the Supreme Court uses a variety of guidelines to determine whether specific kinds of speech are protected by the First Amendment. One important distinction is between private one-on-one conversations (including telephones and mail) and the regulation of mass communications (including newspapers, radio, and television). Social media is tricky because it spans both.

Another important distinction is between government restrictions and actions by private companies, including social media platforms. For example, Twitter permanently suspended Donald Trump's account on January 8, 2021, two days after thousands of his supporters stormed the Capitol Building in Washington, D.C. Trump sued, arguing that this suspension violated his First Amendment rights. A federal judge dismissed the lawsuit, noting that, "The First Amendment applies only to governmental abridgements of speech, and not to alleged abridgements by private companies."

The Comstock Laws are a salient example of the tension between the value of free speech and a desire to protect society.

The Comstock Laws

Anthony Comstock was born in 1844 in a small Connecticut town and grew up in a strictly religious family. He joined the Union Army during the Civil War and was scarred by his fellow soldiers' profanity and other offensive behavior. After the War, Comstock lived in a boardinghouse in New York City, along with other unmarried men who were entertained by drinking, gambling, and loose women. He was so outraged and disgusted that he embarked on a life-long crusade against debauchery.

In 1873 he founded the New York Society for the Suppression of Vice and persuaded the U.S. Congress to pass the Comstock Act, which was intended to protect public morals by making it illegal to send anything that was "obscene, lewd or lascivious," "immoral," or "indecent" through the U.S. mail. The federal government passed several Comstock-related laws and so did twenty-four state legislatures. The Comstock Act also prohibited the sale, gift, or possession

Figure 15.1 Anthony Comstock, vice suppressor.

of obscene writings, pictures, or advertisements—though it did not define "obscene."

Comstock was easily offended. After a White House reception that he had been invited to by President Ulysses S. Grant, he described his revulsion with the way the women were dressed:

They were brazen—dressed extremely silly—enameled faces and powdered hair—low dresses—hair most ridiculous and altogether most extremely disgusting to every lover of pure, noble, modest woman . . . . How can we respect them? They disgrace our land and yet consider themselves ladies.

Comstock was appointed a special agent of the U.S. Postal Service and empowered to carry a gun and arrest people he deigned to have violated the law named after him. He was given a rail pass that allowed him to travel by train, without charge, anywhere in the country in order to, in his words, be "the weeder in God's garden."

In words that might equally well be applied to the Internet today, Comstock wrote that

The mail of the United States is the great thoroughfare of communication leading up into all our homes, schools and colleges. It is the most powerful agent to assist this nefarious business, because it goes everywhere and is secret. It surely needs no argument here to convince the most exacting of all decent men, that no department of Government should be prostituted to serve this infamous traffic, nor become party to it, by continuing to serve these loathsome creatures after the character of their hellish business . . . is known.

In practice, he arrested the author of a pamphlet criticizing marriage, a publisher who distributed this pamphlet through the mail, and a person who was mailed the pamphlet. Comstock also arrested doctors who mailed patients birth control devices or information about pregnancy, contraception, or abortion. He arrested a bookseller who mailed him a copy of Walt Whitman's *Leaves of Grass*. He arrested people who imported art books containing nude paintings. He arrested publishers of medical books showing human anatomy. He arrested people who lectured about birth control. He also went after gamblers, financial scammers, prostitutes, and patent medicine peddlers.

Shortly before he died, he boasted that

In the forty-one years I have been here I have convicted persons enough to fill a passenger train of sixty-one coaches, sixty coaches containing sixty passengers each and the sixty-first almost full. I have destroyed 160 tons of obscene literature.

He also bragged that he had driven fifteen people to commit suicide.

Eventually (not until the 1960s and 1970s!), Comstock restrictions on birth control were declared unconstitutional by the U.S. Supreme Court—though not on First Amendment grounds. In *Griswold v Connecticut* (1965) the Court ruled that restrictions on married couples buying and using contraceptive devices violated the "right to marital privacy." In *Eisenstadt v Baird* (1972) the Court declared that the Equal Protection Clause of the Fourteenth Amendment gave unmarried couples the same privacy rights.

Corporate Baby Steps

Social media companies are currently taking a few baby steps to clean up the garbage they carry and promote. YouTube says that its algorithms will try to avoid recommending conspiracy videos. WhatsApp does not allow messages to be forwarded more than five times. Facebook says that it has switched from an either/or model to a continuous scale in which the closer something is to being prohibited entirely, the less likely it is to be featured by its search and recommendation algorithms. Such approaches create frictions while avoiding First Amendment issues with outright bans. These measures won't stop the blight of mischievous content, but they may slow down the spread of the infection.

In September and October 2021, the *Wall Street Journal* published a series of nine damning articles based on tens of thousands of internal Facebook documents that Frances Haugen, a former Facebook product manager, had given to the Securities and Exchange Commission and the *Journal*. Among the revelations, the *Journal* reported that Facebook executives promoted content that was clearly untrue and inflammatory, knew that Instagram was hurting young girls' self-esteem but did nothing about it, and did little to stop drug trafficking and other illegal activities.

After the explosive *Journal* reports, Haugen testified before Congress and pleaded for government regulation to force Facebook to do what it was unwilling to do on its own.

Government Action

The government regulates telephones, radio, and television, and it can regulate Internet communication too. Balancing the value of free speech and consumer protection need not be entrusted to voluntary actions by Twitter, Instagram, Facebook, YouTube, and other digital platforms. Congress should enact laws that specify clearly the responsibilities of platform operators. Left to make their own decisions, private efforts are likely to be half-hearted. It is easier for companies to do nothing than to risk lawsuits. In addition, courts frequently defer to Congress in ambiguous situations—so Congress should step up and make its intentions clear.

For example, short of shutting down Facebook, Instagram, and the like, the government might raise the minimum age from 13 years to 16 or 18 and require proof of both age and identity, comparable to the identification required to obtain a drivers license, register to vote, or purchase firearms. Social media platforms can be required to satisfy the KYC ("Know Your Customer" or "Know Your Client") regulations that currently apply to most financial institutions and many non-financial companies. KYC regulations specify due diligence rules for verifying a customer's identify and can be done through video streaming and can involve passports, driver's licenses, and other documents in addition to biometric verification.

Overly broad censorship will not survive judicial scrutiny, but narrowly defined restrictions can if the laws are content neutral in the sense that they do not favor some political viewpoints over others.

Just as banks notify law enforcement agencies of suspicious financial transactions, so social media companies can be required to notify police and regulators of terrorist recruitment, planned violence, and other illegal activities, since government agencies are better suited to investigate such activities and enforce the relevant laws and regulations.

The War against Bot Armies

There are two ways to attack Internet falsehoods—via the sender or via the content.

It has been estimated that more than half of all Internet traffic is bots—algorithms that perform automated tasks without human intervention. Bots can do mindless repetitive chores much faster and more carefully than humans—for example, roaming the Web to find search engine content. However, bots can be also used to send spam, break into user accounts, or launch distributed denial of service (DDoS) attacks that overwhelm, slow down, and sometimes shut down online services.

Bots are often programmed to imitate humans. A relatively benign use is performing routine customer service functions. A more malignant purpose is using social media for financial and emotional scams or to spread falsehoods and fictitious conspiracy theories.

One way to hamper malicious bots is to identify and close the fake accounts that they use. Savvy tech companies that use our Internet activities to identify what we like and dislike can surely do a better job of identifying fake accounts. It is admittedly relatively easy for the malevolent to create new accounts in place of closed accounts, but tech companies that can create plagiarism checkers can surely create algorithms that detect content that has been repackaged and resent from new accounts—and purge the content and close the accounts immediately after they are opened.

Bot mitigation measures should also include the widespread distribution of reliable antivirus and bot removal software to combat botnets that send missives from infected computers. If we can create effective COVID-19 vaccines, we can surely create bot shields and cures. The real barrier is that there is no financial incentive to do so.

Anonymity is a tricky issue in that it allows people to speak honestly, but can also be used to hide harmful or illegal activities. A compromise might be to allow users to use pseudonyms, but require that the identity of persons creating accounts be verified.

In addition to closing fake accounts, a second tactic in the war against bots is to make it difficult for bots to access media platforms and to remain on them if they do gain access. One of the first strategies for filtering out bots was an access box with distorted text (Figure 15.2) that is called a CAPTCHA (Completely Automated Public Turing test to tell Computers and Humans Apart). The success of a text CAPTCHA relies on the fact that computer algorithms often struggle to recognize text that they have not been trained on.

Figure 15.2 Prove you are not a robot.

Other CAPTCHAs ask users to click on boxes like those in Figure 15.3 that include images of things like cars, cows, and cats. Computer algorithms based on the mathematical representation of the pixels in images have trouble identifying partial images and images that are fuzzy, distorted, partly obscured, or viewed from odd angles.

As computer algorithms got better at deciphering text and images, Google made its reCAPTCHAs increasingly difficult, so tough that they became challenging for humans as well as algorithms. At the extreme, an impossible test would keep out all bots, but would also keep out all humans.

In 2014, Google introduced the "No CAPTCHA reCAPTCHA." Users are asked to make a single click on a box (as in Figure 15.4), confirming that they

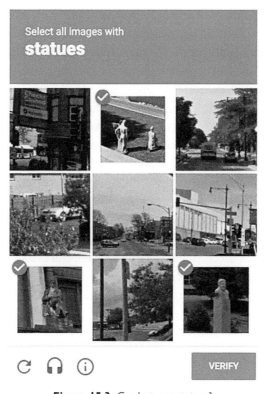

Figure 15.3 Can bots see statues?

are not robots. This simple checkbox seems a step backward from difficult text and image challenges, but it is actually a huge step forward in the battle against bots.

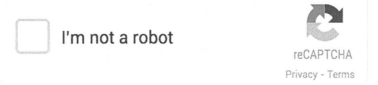

Figure 15.4 Positively human.

Google's secret sauce is not whether the user clicks on the checkbox (computers can do that easily), but what the user does before and immediately after clicking on the checkbox. Google monitors a lot of what we do on our computers and estimates the probability that a user is a human based on the search history, how the mouse is moved, and other proprietary secrets (in the war against bots, the less the bot masters know, the better). For instance, a bot that programmatically clicks a checkbox would not show any mouse movement, let alone the kind of imperfect hand movements made by humans, and might also click the box faster than any human could (the same way that Watson pushed its Jeopardy button faster than its human opponents).

If the Google's No CAPTCHA reCAPTCHA is not certain about whether it is dealing with a human or an algorithm, Google supplements the robot checkbox with traditional CAPTCHA tests.

Bot masters might respond to No CAPTCHA reCAPTCHAs by modifying the code to simulate a wobbly mouse movement and delay the button click—forcing Google engineers to use other human cues. It is a war out there. The point is that there is a war to be fought and the battle to disarm bots ought to include blocking bot access to social media. I am hopeful that Google's engineers can stay a few steps ahead of the bot masters.

Fighting Falsehoods and Lies

In addition to hobbling bots that spread falsehoods and lies, society can fight the untruths directly.

Phony stories generated and spread by people looking to make money can be short-circuited by killing the financial incentive. One step is to limit the forwarding of obviously phony news. A second step is to avoid putting questionable stories at the top of recommended lists or trending lists. A third step is to limit the ability of peddlers of phony stories to purchase ads. A fourth step is to shut down accounts that originate falsehoods.

Tricksters with political intentions are not motivated by advertising dollars and won't be dissuaded by declining revenue. Facebook reduced the spread of disinformation during the 2020 presidential election campaign (showing that it is possible), and then went back to its normal see-no-evil policy afterward—presumably because identifying lies is expensive and hurts advertising revenue. Left to make their own decisions, Facebook and other social media platforms are likely to allow and even feature malignant content if it is profitable to do so—which is why government mandates are needed. Yes, AI algorithms are imperfect and need to be augmented with expensive human fact checkers, but social media platforms can afford to do this and should be required to do it.

Wikipedia has been a spectacular success relying solely on volunteers to write and edit content. An army of public-spirited volunteers could similarly identify phony news stories and use a team of experts to settle disputes. There will inevitably be ambiguities. Germany's 2017 NetzDG law (the "Facebook Act") requires social networks to delete "obviously illegal" content within 24 hours but does not define what is "obviously illegal" and gives a seven-day deadline for non-obvious cases. The European Union's 2022 Digital Services Act requires social media companies to remove content that is illegal, disinformation, or otherwise harmful quickly or face fines equal to up to 6 percent of annual global revenue but is vague about the details of what warrants expulsion.

Media platforms can slow the spread of disinformation by prohibiting forwarding until a post has been fact-checked. Clearly false stories and videos can be tagged; questionable posts can be labeled as possibly false and misleading. In addition to warning labels, media platforms can make false stories and videos harder to find instead of putting them at the top of recommendation lists.

Breaking Filter Bubbles

In addition to creating friction against the dissemination of false news stories, bogus conspiracy theories, and other harmful content, media platforms can enhance the distribution of real news and other authoritative information by putting these near the top of suggested links. Free speech is intended to allow a marketplace of ideas and that remains a worthwhile goal—certainly better than filter bubbles.

Suggesting content that is slightly outside a user's comfort zone can sow seeds of doubt and make it harder for users to find content that is as extreme or more extreme than what they are used to. The adjustments should not be 180°. If the recommended content is too different, challenging, or aggressive, people are likely to reject it as heavy-handed propaganda that reinforces their conspiracy theories about the elites trying to manipulate them.

Forewarned is Forearmed

Another weapon in the war against lies, lies, and more lies would be for schools to teach media literacy, including the ways in which we are manipulated by media and the tricks that advertisers and others use to mislead and misinform.

Many people have trouble simply distinguishing between fact and opinion. If an opinion supports their viewpoint, they are apt to misinterpret it as a fact. If a fact challenges their viewpoint, they are likely to dismiss it as a mistaken opinion. For example, a 2018 Pew Research Center survey of U.S. adults asked respondents to identify whether each of ten statements was "a factual statement (whether accurate or not) or an opinion statement (whether agreed with or not)."

For the factual statement, "President Barack Obama was born in the United States," 37 percent of Republicans and 11 percent of Democrats mislabeled the statement as an opinion. For the opinion, "Increasing the federal minimum wage to $15 an hour is essential for the health of the U.S. economy," 37 percent of Democrats and 17 percent of Republicans mislabeled the statement as factual.

Quantitative literacy is increasingly invaluable for our living amidst the data deluge. We need to recognize how data, statistics, and graphs can be used and abused—and the learning can begin in elementary school and continue through college and even graduate school.

Thinking Backwards

Israel has a very high COVID-19 vaccination rate; yet, on August 15, 2021, 58 percent of the Israelis hospitalized for COVID-19 were fully vaccinated—suggesting that vaccinations are ineffective or even harmful.

This is a great example of the commonplace confusion about inverse probabilities. The chances that a randomly selected player in the English Premier League played soccer as a teenager is 100 percent, but the probability that a randomly selected teenage soccer player will play in the Premier League is virtually zero. In our COVID-19 example, the probability that a person who is vaccinated might be hospitalized for COVOD-19 is not at all the same as the probability that a person hospitalized for COVID-19 was vaccinated.

The calculation of inverse probabilities was formulated by Thomas Bayes, an eighteenth-century mathematician and minister who was evidently trying to calculate the probability of God's existence, given the way the world is, from the inverse probability that the world would be the way it is if there were a God. He didn't have the information he needed to calculate this probability, but his efforts led him to what is now called Bayes' Rule, which is invaluable for calculating inverse probabilities (and more).

For the Israeli COVID-19 data, the relevant data are in Table 15.1.

Of the 515 persons hospitalized for COVID-19, it is true that 58 percent were fully vaccinated: 301/515 = 0.584. Far more relevant is a comparison of the probability that a person who has been vaccinated will be hospitalized with the same probability for people who were not vaccinated. As shown in Table 15.1, these probabilities are

Probability hospitalized if not vaccinated = 214/1,302,912 = 0.00016425

Probability hospitalized if vaccinated = 301/5,634,634 = 0.00005342

The probability of being hospitalized is 3.07 times higher for those who are not vaccinated:

$$\text{risk ratio} = \frac{0.00016425}{0.00005342} = 3.07$$

Even though those who were hospitalized were more likely to be vaccinated, those who were not vaccinated were far more likely to be hospitalized. The vaccine works!

Table 15.2 shows the risk ratios by age. (The risk ratios are infinite for those under the age of 30 because there were no vaccinated hospital patients under the age of 30.)

The proverbial bottom line is that the fact that nearly 60 percent of the people hospitalized for COVID-19 were fully vaccinated does not demonstrate that COVID-19 vaccines are ineffective. These data (which were cited far and wide by anti-vaxxers) actually confirm the effectiveness of COVID-19 vaccines.

Table 15.1 *Probability of Hospitalization if Vaccinated or Not Vaccinated*

	Not fully vaccinated	Fully vaccinated	Total
Hospitalized for COVID-19	214	301	515
Not hospitalized for COVID-19	1,302,698	5,634,333	6,937,031
Total population	1,302,912	5,634,634	6,937,546
Probability hospitalized	0.00016425	0.00005342	
Risk ratio		3.07	

Table 15.2 *Risk Ratios by Age*

Age range	Risk ratio
12–15	∞
16–19	∞
20–9	∞
30–9	31.25
40–9	16.39
50–9	13.89
60–9	8.85
70–9	9.62
80–9	5.29
90+	13.16

The point here is the value of statistical literacy—not just for Bayes' Rule, but for recognizing all sorts of statistical fallacies and mischief.

Understanding and Appreciating Science

Science was supposed to replace superstition and rumors with logic, reason, and empirical evidence. It still can. A starting point for resurrecting the reputation of science is better science education. Memorizing the names of the parts of a cell and then forgetting the names after a test is not scientific understanding. Nor is deciphering the periodic table or memorizing trigonometric formulas. Science is fundamentally about being curious—about how things work and why they sometimes don't work. Richard Feynman's journey to Nobel laureate began with a boyhood curiosity about how radios work. He tinkered with them, took them apart, and put them back together. He fixed other people's radios. He loved it.

He later wrote about his life-long curiosity:

When I was in high school, I'd see water running out of a faucet growing narrower, and wonder if I could figure out what determines that curve. I found it was rather easy to do. I didn't have to do it; it wasn't important for the future of science; somebody else had already done it. That didn't make any difference: I'd invent things and play with things for my own entertainment.

Kids don't have to become Nobel laureates to appreciate how science can satisfy their curiosity. Kids who appreciate science can grow up to respect science and become scientists.

Data Torturing and Data Mining

Data torturing and data mining are conceptually different pitfalls, but the countermeasures are similar.

Data torturing involves manipulating, pruning, and augmenting data to get low p-values—hence the label *p-hacking*. A p-hacker might start off looking for evidence that Chinese-Americans can postpone death until after the Harvest Moon Festival, which is celebrated on the fifteenth day of the eighth moon of the lunar calendar by families gathering for festival meals that include traditional moon cakes eaten outdoors at midnight.

If the initial statistical results are disappointing, the data might be manipulated in several ways—different sexes, different ages, different ways of counting whether a death occurred before or after the celebration. Eventually, a persistent researcher might discover that Chinese-American women 75 years old and older have fewer deaths in the week preceding the festival—if deaths on the festival days are counted as having occurred *after* the festival. (Yes, a paper published in the *Journal of the American Medical Association* reached exactly this conclusion.)

Data mining involves plundering data, with no well-defined purpose in mind other than finding a statistical pattern—hence the label *HARKing*. A researcher might calculate moving averages over periods of one to six weeks of the number of Google searches for 98 different words and discover that the Dow Jones Industrial Average tended to rise when the number of weekly searches for the word *debt* was above its three-week moving average. There were 1,176 strategies to be data-mined so it is no surprise that one strategy predicted the past successfully. (Yes, this, too, is a real paper, co-authored by prominent professors.)

The first hurdle in the battle against p-hacking and HARKing is that some researchers don't realize the pitfalls in these research strategies. Too many view p-hacking and HARKing as virtues, not vices. Thus, the necessary first step for slowing the p-hacking/HARKing express is for researchers to recognize that it is a problem.

Once recognized, there are two ways to counter p-hacking and HARKing. First, journals should de-emphasize statistical significance. Second, replication studies should be rewarded.

De-emphasizing Statistical Significance

P-hacking and HARKing are fundamentally fueled by the belief that statistical significance is a prerequisite for publication. Researchers who p-hack torture data until a desired level of statistical significance is reached. Researchers who HARK believe that their data-mined results are worth reporting if the p-values are sufficiently low. Correlation supersedes causation when p-values are all that matter.

A direct way to fight p-hacking and HARKing is to eliminate the incentive by removing statistical significance as a hurdle for publication. P-values can help us assess the extent to which chance might explain empirical results, but they should not be the primary measure of a model's success. A truly useful model should have reasonable coefficients that allow the model to make plausible predictions. A model that predicts that Asian-Americans are more likely to have heart attacks on the fourth day of every month is not persuasive, no matter how low the p-values.

Artificial thresholds like $p < 0.05$ encourage unsound practices. Many journal editors now recognize this unfortunate incentive and publish papers that are well-done with important results that do not happen to be statistically significant. More journals should do the same.

Making Reproducibility Easy

It shouldn't take a guerrilla army of scientific sleuths to uncover errors in the data and in the mathematical and statistical calculations reported in published papers. All journals should require authors, before publication, to provide all nonconfidential data and computer code for journals to make available online to anyone who is interested. Knowing that it will be easy for others to scrutinize the data and check the calculations, authors will surely be more careful.

Making Replication Routine

When fresh data reveal widely publicized results to be untrustworthy, there is less reason to believe reported scientific results. Neither of the papers discussed at the start of this section has been replicated. The Harvest Moon Festival study flopped when tested with 15 years of additional data. The *debt* strategy flopped when tested with 7 years of additional data.

Detailed descriptions of methodology are often cryptic, which makes replication tests difficult. J. B. Rhine, for example, was notorious for responding to replication failures by claiming that the new tests were done somewhat differently from the way he did them. In this Internet age, it is relatively easy to require authors to provide journals (again, before publication) with a detailed description of the methods used, which can be posted online to facilitate replication tests. If researchers who are tempted to torture or mine data know that their results are likely to be retested, they may be less likely to risk embarrassment by reporting dodgy results.

Replication tests need replicators. Highly skilled researchers are generally enmeshed in their own work and have little reason to spend their time trying to replicate other people's work. One alternative is to make a replication study of an important paper a prerequisite for a PhD or other degree in an empirical field. Such a requirement would allow students to see firsthand how research is done and would also generate thousands of replication tests.

A famous example of this happened in 2012 when Thomas Herndon, a graduate student at the University of Massachusetts Amherst, took a statistics course that required students to replicate a famous research paper. Herndon chose a recent influential paper by two Harvard economists, Carmen Reinhart and Ken Rogoff, that had concluded that a nation is likely to be thrown into an economic recession when the ratio of federal government debt to the nation's annual output exceeds 90 percent.

There is no reason why the level of debt has to be less than annual output. Many homeowners have mortgage balances that exceed their annual income but they are not required to pay off the debt within a year. They might pay off their mortgage over 30 years or they might pay it off when they sell their house. Governments have the additional options of raising taxes or printing money. These aren't necessarily desirable options, but there is no logical reason why debt beyond a certain fixed point should trigger a recession.

Nonetheless, the Reinhart–Rogoff study was cited by fiscal hawks worldwide as evidence that they should raise taxes and cut government spending to reduce government debt even if these actions caused an economic recession—which they did. Yes, they caused recessions in the name of avoiding recessions.

Herndon gathered the same data that Reinhart and Rogoff ostensibly used and did not get the same results they got. He and his professors figured that he must have done something wrong but they couldn't identify any errors. Herndon contacted Reinhart and Rogoff but the famous Harvard professors ignored the pleas of an unknown graduate student. Herndon persisted and he finally obtained their data and spreadsheet calculations and soon discovered the problem—actually several problems involving computational errors and questionable assumptions. (A fuller discussion is in my book *Standard Deviations: Flawed Assumptions, Tortured Data, and Other Ways to Lie with Statistics*.)

The point here is that Herndon should not have had to struggle so mightily to obtain the data and methodology underlying the Reinhart–Rogoff paper. Even better, if Reinhart and Rogoff knew that a bulldog like Herndon was going to take a close look, they might have been more diligent.

A requirement that authors make their data and methodology publicly available is invaluable; however, the Reinhart–Rogoff incident points out an unfortunate barrier for some replication studies. Herndon found errors in the original analysis done by Reinhart and Rogoff but what if the only errors had been torturing the data for statistical significance? Reinhart and Rogoff looked at data for the years 1946–2009. Any attempt to replicate their study with fresh data would have had to wait several years, even decades, to accumulate enough data—by which time their policy recommendation would have caused even more irreparable harm.

Pre-registration

Another protection against data torturing and data mining is a requirement that researchers file a detailed description of their research plans before they begin their research. A study of whether people can postpone death until after the Harvest Moon Festival would be far less likely to reach an erroneous conclusion if the sex and ages of the people being considered were identified in advance. A study of whether a trending keyword can predict stock prices would be far less likely to reach an erroneous conclusion if the keyword was specified beforehand.

Another benefit is that journals may be more willing to publish a paper with p-values above 0.05 if the paper is pre-registered. It is worth knowing that there is no persuasive evidence that people can postpone death until after a celebration or that keyword trends can be used to predict stock prices.

Relatively few journals currently require pre-registration—perhaps because it so easy to game the system: collect the data, torture or mine the data to obtain interesting results, and then file a pre-plan that does not reveal that the study has already been completed.

Such gamesmanship is difficult when large-scale experiments are done, especially if there are several researchers involved who might see and report the misconduct. In Brian Wansink's pizza studies discussed in Chapter 5, a research assistant saw the mischief first hand and later wrote, "I remember him saying it so clearly: 'Just keep messing with the data until you find something,'" It is risky to file a pre-plan after a study is done if there are witnesses to the shenanigans.

Unfortunately, such gamesmanship is relatively easy when a single researcher is scrutinizing observational data. If Wansink had done the study on his own and not boasted about his p-hacking and HARKing afterward, he might have gotten away with it.

Andrew Gelman, a very careful and scrupulous statistician, has noted three other problems with pre-registration. First, researchers may already be very familiar with the data they are going to use and the analyses others have done, so they implicitly p-hack without recognizing that they are p-hacking. Second, pre-registration plans will inevitably omit some unanticipated forking paths. In the full-moon motorcycle study, the researchers did not realize ahead of time that there were time-zone ambiguities. Third, sometimes the data suggest plausible theories that can be tested with fresh data. Gelman wrote that

In our many applied research projects, we have learned so much by looking at the data. Our most important hypotheses could never have been formulated ahead of time.

On the other hand, a study can be improved by suggestions solicited in a peer-reviewed pre-plan. The editors in charge of the CRSP replication studies of power posing wrote of the peer-reviewed pre-registration:

Without exception, the method, design, or analysis of every proposed study was modified in some way following the initial Stage I review. True experts in the fields of embodied cognition, hormones, and other relevant areas of expertise provided advice to researchers before they spent precious resources conducting these studies. . . . Reviewers were not on the lookout for the many ways the researchers failed to do what they should have (often a reflection of reviewers trying to show how smart they are) but instead approached these proposals with the mindset of, "What would I do to make this the best research possible?" Refreshing indeed.

Final Thoughts

Here is a summary of several possible actions. I am hopeful that some might be implemented by the time you read this, and that more are on the way.

Disinformation

1. Raise the minimum age for social media accounts from 13 years old to 16 or 18.

2. Require people creating Internet accounts to be verified through clear evidence of identity.

3. Block bot access to social media.

4. Require social media companies to notify police and regulators of terrorist recruitment, planned violence, and other illegal activities.

5. Use humans and algorithms to identify phony news stories and videos, with a team of experts used to settle disputes.

6. Tag stories and videos that are clearly false.

7. Label questionable posts as possibly false and misleading and provide links to authoritative information.

8. Prohibit forwarding until a post has been reviewed and fact-checked.

9. Posts that have been forwarded twice cannot be forwarded again unless the user copies and pastes the original post.

10. Shut down accounts that initiate false news stories.

11. Limit the ability of peddlers of phony stories to purchase ads.

12. Do not put false stories and videos on recommended lists or trending lists.

13. Put authoritative information that is less extreme, but not polar opposite to users' beliefs, at the top of recommendation lists.

14. There should be more (and better) courses on media, quantitative, and scientific literacy, both online and in schools.

Data Torturing and Data Mining

1. Authors should be encouraged to report p-values instead of yes/no labels regarding statistical significance.

2. Authors should be encouraged to report confidence intervals.

3. Authors should discuss the plausibility of their estimated coefficients.

4. Authors should discuss whether their models' estimated coefficients are substantial—have oomph.

5. Journal editors should not require p-values below an artificial threshold as a prerequisite for publication.

6. Journal editors should prioritize papers that file a detailed description of the research plans before beginning the research.

7. Journals should not publish empirical research until nonconfidential data and a detailed description of the methods are made publicly available.

8. Reproducibility and replication studies should be encouraged and rewarded—perhaps by foundations or government agencies.

9. A reproducibility or replication study of an important paper should be a prerequisite for a PhD or other degree in an empirical field.

10. Courses in statistical literacy and reasoning should be an integral part of school curricula and made available online, too.

11. Statistics courses in all disciplines should include substantial discussion of Bayesian methods.

12. Statistics courses in all disciplines should include substantial discussion of p-hacking and HARKing.

Many actions intended to combat disinformation will require the cooperation and expertise of tech companies, but they need to be motivated by government laws, regulations, and oversight. The suggestions intended to reduce p-hacking and HARKing rely on the informed goodwill of researchers and journal editors.

It will be messy, but there aren't any easy, effective options and we desperately need to restore the luster of science and scientists.

REFERENCES

Abutaleb, Yasmeen, McGinley, Laurie, & Johnson, Carolyn Y. 2020. How the "deep state" scientists vilified by Trump helped him deliver an unprecedented achievement. *Washington Post*, December 14.

Agüera y Arcas, Blais. 2021. Do large language models understand us? Medium, December 16.

Alan Turing Institute. 2021. Data science and AI in the age of COVID-19.

Allyn, Bobby. 2020. Researchers: Nearly half of accounts tweeting about coronavirus are likely bots. NPR, May 20.

Ambrosino, Brandon. 2016. Facebook is a growing and unstoppable digital graveyard. BBC Future, March 13.

Anderson, Rick. 2014. The scam, the sting, and the reaction: Labbé, Bohannon, Sokal. The Scholarly Kitchen, February 28.

Anderson, T. W., Suranyi, G., & Beaton, G. H. 1974. The effect on winter illness of large doses of vitamin C. *Canadian Medical Association Journal*, 111(1), 31–36.

Bakker, Marjan, & Wicherts, Jelte M. 2011. The (mis)reporting of statistical results in psychology journals. *Behavior Research Methods*, 43(3), 666–678.

Ball, P. 2005. Computer conference welcomes gobbledegook paper. *Nature*, 434(7036), 946.

Benford, Frank. 1938. The law of anomalous numbers. *Proceedings of the American Philosophical Society*, 78(4), 551–572.

Berger, J. M. 2014. How ISIS games Twitter. Atlantic, June 16.

Braeunig, Robert A. 2006. Did we land on the Moon? Rocket and Space Technology.

Brembs, Björn. 2018. Prestigious science journals struggle to reach even average reliability. *Frontiers in Human Neuroscience*, 12 (37).

Brembs, Björn, Button, Katherine, & Munafò, Marcus. 2013. Deep impact: Unintended consequences of journal rank. *Frontiers in Human Neuroscience*, 7, 291.

Brennan, L., Watson, M., Klaber, R., & Charles, T. 2012. The importance of knowing context of hospital episode statistics when reconfiguring the NHS. *British Medical Journal*, 344.

Brin, Sergey, & Page, Lawrence. 1998. The anatomy of a large-scale hypertextual web search engine. *Computer Networks*, 30, 107–117.

Brown, Nicolas J. L. 2019. An update on our examination of the research of Dr. Nicolas Guéguen. Nick Brown's Blog, May 19.

Brown, Nicolas J. L. 2021. Some problems in the dataset of a large study of Ivermectin for the treatment of Covid-19. Nick Brown's Blog, July 15.

Brown, Nicolas J. L., Sokal, Alan D., & Friedman, Harris L. 2013. The complex dynamics of wishful thinking: The critical positivity ratio. *American Psychologist*, 68(9), 801–813.

Bump, Philip. 2013. 12 Million Americans Believe Lizard People Run Our Country. The Atlantic, April 13.

Cabanac, Guillaume, & Labbé, Cyril. 2021. Prevalence of nonsensical algorithmically generated papers in the scientific literature. *Journal of the Association for Information Science and Technology*, 72(12), 1461–1476.

Cafbajal, Erica. 2021. Nearly 60% of hospitalized COVID-19 patients in Israel fully vaccinated, data shows. Becker's Hospital Review, August 19.

Cahill, Aoife, Chodorow, Martin, & Flor, Michael. 2018. Developing an e-rater advisory to detect Babel-generated essays. *Journal of Writing Analytics*, 2, 203–224.

Camerer, Colin F., Dreber, Anna, Forsell, Eskil, Ho, Teck-Hua, Huber, Jürgen, Johannesson, Magnus, et al. 2016. Evaluating replicability of laboratory experiments in economics. *Science*, 351, 1433–1436.

Camerer, Colin F., Dreber, Anna, Holzmeister, Felix, Ho, Teck-Hua, Huber, Jürgen, Johannesson, Magnus, et al. 2018. Evaluating the replicability of social science experiments in *Nature* and *Science* between 2010 and 2015. *Nature Human Behaviour*, 2(9), 637–644.

Carney, Dana R. (nd). My position on "Power Poses". https://faculty.haas.berkeley.edu/dana_carney/pdf_my%20position%20on%20power%20poses.pdf.

Carney, Dana R., Cuddy, Amy J.C., & Yap, Andy J. 2010. Power posing: brief nonverbal displays affect neuroendocrine levels and risk tolerance. *Psychological Science*, 21(10), 1363–1368.

Cesario, Joseph, Jonas, Kai J., & Carney, Dana R. 2017. CRSP special issue on power poses: what was the point and what did we learn? *Comprehensive Results in Social Psychology*, 2(1).

Chapman, Cath, & Slade, Tim. 2015. Rejection of rejection: a novel approach to overcoming barriers to publication. *British Medical Journal*, 351.

Chen, Adrian. 2015. The agency. *New York Times*, June 2.

Clark, Patrick. 2019. Zillow wants to flip your house. Bloomberg, February 14.

Clark, Patrick. 2021. Zillow shuts home-flipping business after racking up losses. *Los Angeles Times*, November 2.

CNBC. 2021. Breaking News. CNBC, March 8.

Cohen, Elizabeth. 2021. CDC scientists recall "death by a thousand cuts" as they try to rebuild the agency's reputation. CNN, February 5.

Cokol, Murat, Iossifov, Ivan, Rodriguez-Esteban, Raul, & Rzhetsky, Andrey. 2007. How many scientific papers should be retracted? *EMBO Reports*, 8(5), 422–423.

Colyer, Jeff, & Hinthorn, Daniel. 2020. These drugs are helping our coronavirus patients. *Wall Street Journal*, March 20.

Crockett, Zachary. 2017. How the "King of Fake News" built his empire. The Hustle, November 7.

Croucher, Shane. 2019. Donald Trump pretended to be a corporate raider, talking up stock prices before quietly selling his shares: Report. *Newsweek*, May 8.

Crumbaugh, James. 1966. A scientific critique of parapsychology. *International Journal of Neuropsychiatry*, 2(5), 523–531.

Cuddy, Amy J. C., Norton, Michael I., & Fiske, Susan T. 2005. This old stereotype: The pervasiveness and persistence of the elderly stereotype. *Journal of Social Issues*, 61(2), 267–285.

Cvetkovska, Saska, Belford, Aubrey, Silverman, Craig, & Feder, Lester. 2018. The secret players behind Macedonia's fake news sites. Organized Crime and Corruption Reporting Project, July 18.

Dadkhah, Medhi, Maliszewski, Tomasz, & da Silva, Jaine A. Teixeira. 2016. Hijacked journals, hijacked web-sites, journal phishing, misleading metrics, and predatory publishing: actual and potential threats to academic integrity and publishing ethics. *Forensic Science, Medicine, and Pathology*, 12(3), 353–362.

DeGrave, Alex, Janizek, Joseph, & Lee, Su-In. 2021. AI for radiographic COVID-19 detection selects shortcuts over signal. Nature Machine Intelligence, May.

Delgado López-Cózar, E., Robinson-García, N., & Torres-Salinas, D. 2014. The Google Scholar experiment: How to index false papers and manipulate bibliometric indicators. *Journal of the American Society for Information Science and Technology*, 65(3), 446–454.

Dewey, Caitlin. 2016. Facebook fake-news writer: "I think Donald Trump is in the White House because of me." *Washington Post*, November 17.

Dhar, Rohin. 2016. The trade of the century: When George Soros broke the British pound. Priceonomics, June 17.

Diamond, Dan. 2021. Feuds, fibs and finger-pointing: Trump officials say coronavirus response was worse than known. *Washington Post*, March 29.

Dowd, Gregory Evans. 2015. *Groundless: Rumors, Legends, and Hoaxes on the Early American Frontier*. Baltimore: Johns Hopkins University Press.

Dreber, Anna, Pheiffer, Thomas, Almenberg, Johan, Isaksson, Siri, Wilson, Brad, Chen, Yiling, et al. 2015. Prediction markets in science. *Proceedings of the National Academy of Sciences*, 112(50) 15343–15347.

Duckett, Chris. 2021. Apple CEO sounds warning of algorithms pushing society towards catastrophe. ZDNet, January 29.

Einzig, Paul. 1966. *Primitive Money*, second edition. Oxford: Pergamon Press.

Elgazzar, A., Eltaweel, A., Youssef, S. A., Hany, B., Hafez, M., & Moussa, H. 2020. Efficacy and safety of Ivermectin for treatment and prophylaxis of COVID-19 pandemic. Research Square, 100956.

Emproto, Robert. 2020. Maverick modeller Helmut Norpoth predicts another win for Trump. Stony Brook University News, August 3.

Errington, Timothy M., Mathur, Maya, Soderberg, Courtney K., Denis, Alexandria, Perfito, Nicole, Iorns, Elizabeth, et al. 2021. Investigating the replicability of preclinical cancer biology. *eLife*, 10:e71601.

Fang, Ferric C., & Casadevall, Arturo. 2011. Retracted science and the retraction index. *Infection and immunity*, 79(10), 3855–3859.

Feld, Harold. 2019. *The Case for the Digital Platform Act: Market Structure and Regulation of Digital Platforms*. Roosevelt Institute.

Felton, Edward, Raj, Manav, & Seamans, Robert. 2021. Occupational, industry, and geographic exposure to artificial intelligence: A novel dataset and its potential uses. *Strategic Management Journal*, 42(12), 2195–2217.

Ferguson, Cat, Marcus, Adam, & Oransky, Ivan. 2014. Publishing: The peer-review scam. *Nature*, 515(7528), 480–482.

Feynman, Richard. 1969. What is science? *The Physics Teacher*, 7(6), 313–320.

Fiske, Sjusan T. 2016. A call to change science's culture of shaming. *APS Observer*, 29(9).

Fredrickson, Barbara L., & Losada, Marcial F. 2005. Positive affect and the complex dynamics of human flourishing. *The American Psychologist*, 60(7), 678–686.

Fukuyama, Francis. 2007. The emergence of a post-fact world. Project Syndicate, January 12.

Galak, Jeff, LeBoeuf, Robyn A, Nelson, Leof D, & Simmons, Joseph P. 2012. Correcting the past: failures to replicate ψ. *Journal of Personality and Social Psychology*, 103(6), 933–948.

Galer, Sophia Smith. 2020. The accidental invention of the Illuminati conspiracy. BBC Future, July 11.

Gelman, Andrew. 2016. What has happened down here is the winds have changed. Statistical Modeling, Causal Inference, and Social Science, September 21.

Gelman, Andrew. 2017. Pizzagate, or the curious incident of the researcher in response to people pointing out 150 errors in four of his papers. Statistical Modeling, Causal Inference, and Social Science, February 3.

Gelman, Andrew, & Guzey, Alexey. 2020. Statistics as squid ink: How prominent researchers can get away with misrepresenting data. *Chance*, 25–27.

Ghidina, Marcia J. 2019, Finding god in grain: Crop circles, rationality, and the construction of spiritual experience. *Symbolic Interaction*, 42(2), 278–300.

Gideon, M. K. 2021. Is ivermectin for Covid-19 based on fraudulent research? Gideon M-K; Health Nerd, July 15.

Godwin, Richard. 2019. One giant . . . lie? Why so many people still think the moon landings were faked. *The Guardian*, July 10.

Goldman, Adam. 2019. The Comet Ping Pong gunman answers our reporter's questions. *New York Times*, December 7.

Gottman, John M., Coan, James, Carrere, Sybil, & Swanson, Catherine. 1998. Predicting marital happiness and stability from newlywed interactions. *Journal of Marriage and the Family*, 60(1), 5–22.

Gross, Terry. 2021. How an anti-vice crusader sabotaged the early birth control movement. NPR, July 7.

Guéguen, Nicolas. 2015. Women's hairstyle and men's behavior: A field experiment. *Scandinavian Journal of Psychology*, 56(6), 637–640.

Guzey, A. 2019. Matthew Walker's Why We Sleep is riddled with scientific and factual errors. Alexey Guzey blog.

Hamrick, J. T., Rouhi, Farhang, Mukherjee, Arghya, Feder, Amir, Gandal, Neil, Moore, Tyler, et al. 2021. An examination of the cryptocurrency "pump and dump" ecosystem. *Information Processing & Management*, 58(4), 102506.

Hao, Karen. 2020. AI pioneer Geoff Hinton: "Deep learning is going to be able to do everything." MIT Technology Review, November 23.

Haq, A. U., Zeb, A., Lei, Z., & Zhang, D. 2021. Forecasting daily stock trend using multi-filter feature selection and deep learning. *Expert Systems with Applications*, 168, 114444.

Harrington, Hugh T. 2014. Propaganda warfare: Benjamin Franklin fakes a newspaper. *Journal of the American Revolution*, November 10.

Heaney, Katie. 2015. Meet the croppies. Pacific Standard, October 3.

Heathers, James. 2021. The real scandal about ivermectin. The Atlantic, October 23.

Heaven, Will Douglas. 2021. Hundreds of AI tools have been built to catch covid: None of them helped. MIT Technology Review, July 30.

Herrera-Perez, D., Haslam, A., Crain, T., Gill, J., Livingston, C., Kaestner, V., et al. 2019. Meta-research: A comprehensive review of randomized clinical trials in three medical journals reveals 396 medical reversals. *Elife*. 8, e45183.

Heyman, Richard E., & Smith, Amy M. Slep. 2001. The hazards of predicting divorce without crossvalidation. *Journal of Marriage and the Family*, 63 (2), 473–479.

Higgins, Andrew, McIntire, Mike, & Dance, Gabriel J.x. 2016. Inside a fake news sausage factory: "This is all about income." *New York Times*, November 25.

Hill, R. G., Jr, Sears, L. M., & Melanson, S. W. 2013. 4000 clicks: A productivity analysis of electronic medical records in a community hospital ED. *American Journal Emergency Medicine*, 31(11), 1591–1594.

Hinton, Geoff. 2016. Geoff Hinton: On radiology. Creative Destruction Lab, November 24.

Honavar, Santosh G. 2020. Electronic medical records—The good, the bad and the ugly. *Indian Journal of Ophthalmology*, 68(3), 417–418.

Horizons. 2017. Horizons ETFs launches Canada's first ETF driven by A.I. Press release, November 1.

Horwitz, Jeff. 2021. Facebook says its rules apply to all. Company documents reveal a secret elite that's exempt. *Wall Street Journal*, September 13.

Horwitz, Jeff. 2021. The Facebook whistleblower, Frances Haugen, says she wants to fix the company, not harm it. *Wall Street Journal*, October 3.

Huang, Pien. 2020. Past CDC director urges current one to stand up to Trump. NPR, October 8.

Hughes, H., & Waismel-Manor, I. 2021. The Macedonian fake news industry and the 2016 US election. *PS: Political Science & Politics*, 54(1), 19–23.

Inzlicht, Michael. 2016. Reckoning with the past, February 29.

Ioannidis, John A. 2005. Contradicted and initially stronger effects in highly cited clinical research. *Journal of the American Medical Association*, 294(2), 218–228.

Ioannidis, J. P. A. 2005. Why most published research findings are false. *PLoS Medicine*, 2(8), e124.

Jaffe, Thomas, & Machan, Dyan. 1992. How the market overwhelmed the central banks. *Forbes*, November 9.

John, Leslie K., Loewenstein, George, & Prelec, Drazen. 2012. Measuring the prevalence of questionable research practices with incentives for truth telling. *Psychological Science*, 23(5), 524–532.

Johnson, Eric. 2017. Barack Obama conspiracy theories, brought to you by Google Home. Vox, March 5.

Jones, C. W., Handler, L., Crowell, K. E., Keil, L. G., Weaver, M. A., Platts-Mills, T. F., et al. 2013. Non-publication of large randomized clinical trials: cross sectional analysis. *British Medical Journal*, 347.

Kaempffert, W. 1937. The Duke experiments in extra-sensory perception. *New York Times*, October 10.

Kan, Michael. 2018. Tim Cook: Our data is being "weaponized against us." PC, October 24.

Kato, Hirotaka, Jena, Anupam B., Newhouse, Ruth L., & Tsugawa, Yusuke. 2020. Patient mortality after surgery on the surgeon's birthday: observational study. *British Medical Journal*, 371.

Keaney, John J., Groarke, John D., Galvin, Zita, McGorrian, Catherine, McCann, Hugh A., Sugrue, Declan, et al. 2013. The Brady Bunch? New evidence for nominative determinism in patients' health: retrospective, population based cohort study. *British Medical Journal*, 347.

Kelotra, Amit, & Pandey, Prateek. 2020. Stock Market Prediction Using Optimized Deep-ConvLSTM Model. *Big Data*, 8(1).

Kirby, Emma Jane. 2017. The city getting rich from fake news. BBC News, December 5.

Krueger, Thomas M., & Kennedy, William F. 1990. An examination of the Super Bowl Stock Market Predictor. *Journal of Finance*, 45(2), 691–697.

Krugman, Paul. 2018. Bubble, bubble, fraud and trouble. *New York Times*, January 29.

Kunzmann, Kevin. 2018. Why Are EMRs So Terrible? HCP Live.

Labbé, C. 2010. Ike Antkare, one of the greatest stars in the scientific firmament. *ISSI Newsletter*, 6(1), 48–52.

Labbé, Cyril, & Labbé, Dominique. 2013. Duplicate and fake publications in the scientific literature: how many SCIgen papers in computer science? *Scientometrics* 94, 379–396.

Lagakis, Paraskevas, & Demetriadis, Stavros. 2021. Automated essay scoring: A review of the field. 2021 International Conference on Computer, Information and Telecommunication Systems (CITS), 1–6.

Lane, A., Luminet, O., Nave, G., & Mikolajczak, M. 2016. Is there publication bias in behavior intranasal oxytocin research on humans? Opening the file drawer of one laboratory. *Journal of Neuroendocrinology*, 28(4).

Lane, Síle. 2013. Why the data on all drug trials must be released. *The Guardian*, September 17.

Lee, Stephanie M. 2017. Here's how a controversial study about kids and cookies turned out to be wrong—and wrong again. Buzzfeed News, October 18.

Lee, Stephanie. 2018. Here's How Cornell scientist Brian Wansink turned shoddy data into viral studies about how we eat. Buzzfeed News, February 25.

Leibovic, Leonard. 2001. Effects of remote, retroactive intercessory prayer on outcomes in patients with bloodstream infection: randomised controlled trial. *British Medical Journal*, 323(7327),1450–1451.

Leonard, Devin. 2016. The life and times of a true American moral hysteric. Literacy Hub, May 2.

Liu, Yukun, & Tsyvinski, Aleh. 2018. Risks and returns of cryptocurrency. NBER Working Paper 24877.

Loecher, Markus. 2020. Trouble with TED. Code and Stats, September 3.

McCloskey, D. N. 1985. *The Rhetoric of Economics*. Madison: University of Wisconsin Press.

McCloskey, D. N., & Ziliak, ST. 1996. The standard error of regressions. *Journal of Economic Literature*, 34(1), 97–114.

McCormick, John. 2021. Potential IBM Watson Health sale puts focus on data challenges. *Wall Street Journal*, February 24.

McRae, Mike. 2019. Science's "replication crisis" has reached even the most respectable journals, report shows. ScienceAlert, September 29.

Maguolo, Gianluca, & Nanni, Loris. 2021. A critic evaluation of methods for COVID-19 automatic detection from X-ray images. *Information Fusion*, 76, 1–7.

Marcus, Adam, & Oransky, Ivan. 2018. Meet the "data thugs" out to expose shoddy and questionable research. Science, February 14.

Marcus, Gary, & Davis, Ernest. 2020. GPT-3, Bloviator: OpenAI's language generator has no idea what it's talking about. MIT Technology Review, August 22.

Markoff, John. 2016. Automated pro-Trump bots overwhelmed pro-Clinton messages, researchers say. *New York Times*, November 18.

Mathews, Fiona, Johnson, Paul J., & Neil, Andrew. 2008. You are what your mother eats: evidence for maternal preconception diet influencing foetal sex in humans. *Royal Society: Proceedings: Biological Sciences*, 275(1643), 1661–1668.

Metz, Rachel. 2021. Lemonade: This $5 billion insurance company likes to talk up its AI. Now it's in a mess over it. CNN, May 28.

Meyer, Robinson. 2016. War goes viral: How social media is being weaponized across the world. Atlantic, October 18.

Miller, Chance. 2017. Fake news in featured snippets, and thereby Google Home, is clearly still a problem. 9to5Google, March 6.

Mitchell, Amy, Gottfried, Jeffrey, Barthel, Michael, & Sumida, Nami. 2018. *Distinguishing Between Factual and Opinion Statements in News*. Pew Research Center, June 18.

Morrison, Sara. 2021. A disturbing, viral Twitter thread reveals how AI-powered insurance can go wrong. VOX, May 27.

Mossman, Kate. 2017. Ghostwatch: the Halloween hoax that changed the language of television. NewStatesman, October 19.

Murphy, Tim. 2020. Robert Redfield's epic COVID failure is not a surprise to many HIV and public health experts. The BodyPro, September 28.

Myers, Jolie, & Evstatieva, Monika. 2018. Meet the activist who uncovered the Russian troll factory named in the Mueller Probe. NPR, March 15.

National Academies of Sciences, Engineering, and Medicine. 2019. *Taking Action Against Clinician Burnout: A Systems Approach to Professional Well-Being*. Washington, DC: National Academies Press.

Newcomb, Simon. 1881. Note on the frequency of use of the different digits in natural numbers. *American Journal of Mathematics*, 4(1), 39–40.

Norpoth, Helmut, Undated, A Perfect Storm.

Nosek, Brian A., Cohoon, Johanna, Kidwell, Mallory, & Spies, Jeffrey Robert. 2015. Estimating the reproducibility of psychological science. *Science*, 349 (6251).

Nuijten, Michèle B., Hartgerink, Chris H. J., van Assen, Marcel A. L. M., Epskamp, Sacha, & Wicherts, Jelte. 2016. The prevalence of statistical reporting errors in psychology (1985–2013). *Behavior Research Methods*, 48(4), 1205–1226.

Oliver, J. Eric, & Wood Thomas J. 2014. Conspiracy theories and the paranoid style(s) of mass opinion. *American Journal of Political Science*, 58 (4), 952–966.

Open Science Collaboration. 2015. Estimating the reproducibility of psychological science. *Science* 349, aac4716.

Packer, Milton. 2019. What did the Apple heart study really find? MedPage Today, March 20.

Panesar, Nirmal S., Chan, Noel C. Y., Li, Shi N., Lo, Joyce K Y., Wong, Vivien W. Y., Yang, Isaac B., et al. 2003. Is four a deadly number for the Chinese? *Medical Journal of Australia*, 179(11), 656–658.

Pashler, Harold, & Wagenmakers, Eric. 2012. Editors' introduction to the special section on replicability in psychological science: A crisis of confidence?. *Perspectives on Psychological Science*, 7(6), 528–530.

Pearson, Dave. 2021. FDA going soft on AI reviews? Business Intelligence, January 19.

Perelman, Les. 2020. The BABEL Generator and E-Rater: 21st century writing constructs and automated essay scoring (AES). *Journal of Writing Assessment*, 13(1).

Perry, Tekla S. 2021. Andrew Ng X-Rays the AI Hype. IEEE Spectrum, May 3.

Pesce, Nicole Lyn. 2020. About half of the Twitter accounts calling for reopening America are bots: report. MarketWatch, May 26.

Pickard, Alex. 2012. *Bitcoin: Magic Internet Money*. Research Affiliates.

Pine, Art. 1984. Fixed assets, or: Why a loan in Yap is hard to roll over. *Wall Street Journal*, March 29.

Playboy Advisor. 1969. Letters. *Playboy*, April, p. 62.

Portes, Jonathan. 2012. Sterling: My part in its downfall. Juncture. *The Journal of the Institute for Public Policy Research*, September 17.

Preis, Tobias, Moat, Helen Susannah, & Stanley, H. Eugene. 2013. Quantifying trading behavior in financial markets using Google trends. *Scientific Reports*, 3: 1684.

Prier, Jarred. 2017. Commanding the trend: Social media as information warfare. *Strategic Studies Quarterly*, 11(4), 50–85.

Randall, David, & Welser, Christopher. 2018. The irreproducibility crisis of modern science. National Association of Scholars.

Randi, James. 1983. The Project Alpha experiment: Part 1: The first two years, and Part 2: Beyond the laboratory. *Skeptical Inquirer*. 7-8.

Ranehill, Eva, Dreber, Anna, Johannesson, Magnus, Leiberg, Susanne, Sul, Sunhae, & Weber, Roberto A. 2015. Assessing the robustness of power posing: No effect on hormones and risk tolerance in a large sample of men and women. *Psychological Science*, 26(5) 653–656.

Redelmeier, Donald A., & Shafir, Eldar. 2017. The full moon and motorcycle related mortality: population based double control study. *British Medical Journal*, 359, j5367.

Reeves, Margaret, & Rhine, J. B. 1943. The PK effect: A study in declines. *Journal of Parapsychology*, 7, 76–93.

Reisinger, Don. 2016. How Google is quietly benefiting from Pokémon Go's success. *Fortune*, July 12.

Rhine, J. B. 1935. *Extra-Sensory Perception*, Boston, MA: Bruce Humphries.

Rhine, J. B. 1974. A new case of experimenter unreliability. *Journal of Parapsychology*, 38, 137–153.

Rhine, J. B., & Rhine, Louisa E. 1927. One evening's observation on the Margery mediumship. *Journal of Abnormal and Social Psychology*, 21(4), 401–421.

Robb, Amanda. 2017. Anatomy of a fake news scandal. Rolling Stone, November 16.

Roberts, M., Driggs, D., Thorpe, M., Gilbey, J., Yeung, M., Ursprung, M., et al. 2012. Common pitfalls and recommendations for using machine learning to detect and prognosticate for COVID-19 using chest radiographs and CT scans. *Natural Machine Intelligence* 3, 199–217.

Romero-Brufau, Santiago, Wyatt, Kirk D., Boyum, Patricia, Mickelson, Mindy, Moore, Matthew, & Cognetta-Rieke, Cheristi. 2020. A lesson in implementation: A pre-post study of providers' experience with artificial intelligence-based clinical decision support. *International Journal of Medical Informatics*, 137, 104072.

Rosenthal, Robert. 1979. The "file drawer problem" and tolerance for null results. *Psychological Bulletin*, 86(3), 638–641.

Ross, Casey, & Swetlitz, Ike. 2018. IBM's Watson supercomputer recommended "unsafe and incorrect" cancer treatments, internal documents show. STAT, July 25.

Sallis, Robert, Young, Deborah Rohm, Tartof, Sara Y., Sallis, James F., Sall, Jeevan, Li, Qiaowu, et al. 2021. Physical inactivity is associated with a higher risk for severe COVID-19 outcomes: a study in 48 440 adult patients. *British Journal of Sports Medicine*, 55, 1099–1105.

Sandhu, A., Seth, M., & Gurm, H. S. 2014. Daylight savings time and myocardial infarction. *Open Heart*, 1: e000019.

Scanlon, T. J., Luben, R. N., Scanlon, F. L., & Singleton, N. 1993. Is Friday the 13th bad for your health? *British Medical Journal*, 307(6919), 1584–1586.

Scheck, Justin, Purnell, Newley, & Horwitz, Jeff. 2021. Facebook employees flag drug cartels and human traffickers. The company's response is weak, documents show. *Wall Street Journal*, September 16.

Schreiber, Daniel. undated. Lemonade.

Science Insider. 2015. Hoax-detecting software spots fake papers. Communications of the ACM, March 30.

Serra-Garcia, Marta, & Gneezy, Uri. 2021. Nonreplicable publications are cited more than replicable ones. *Science Advances*, 7(21).

Sides, John. 2015. Fifty percent of Americans believe in some conspiracy theory. Here's why. *Washington Post*, February 19.

Silverman, Craig. 2016. This analysis shows how viral fake election news stories outperformed real news on Facebook. BuzzFeed News, November 16.

Simmons, Joseph P., Nelson, Leif D., & Simonsohn, Uri. 2011. False-positive psychology: undisclosed flexibility in data collection and analysis allows presenting anything as significant. *Psychological Science*, 22, 1359–1366.

Simmons, Joseph P., Nelson, Leif D., & Simonsohn, Uri. 2018. False-positive citations. *Perspectives on Psychological Science*, 13(2), 255–259.

Singal, Jessie. 2017. A popular diet-science lab has been publishing really shoddy research. *New York*, February 8.

Smith, Allan. 2021. Birx recalls "very difficult" call with Trump, says hundreds of thousands of Covid deaths were preventable. NBC News, March 28.

Smith, Gary. 2002. Scared to death? *British Medical Journal*, 325, 1442–1443.

Smith, Gary. 2014. *Standard Deviations: Flawed Assumptions, Tortured Data, and Other Ways to Lie With Statistics*. New York: Overlook; London: Duckworth.

Smith, Gary. 2016. Hurricane names: A bunch of hot air? *Weather and Climate Extremes*, 12, 80–84.

Smith, Gary. 2018. Step away from stepwise. *Journal of Big Data*, 5: 32.

Smith, Gary. 2018. *The AI Delusion*. Oxford: Oxford University Press.

Smith, Gary. 2020. Data mining fool's gold. *Journal of Information Technology*, 35(3), 182–194.

Smith, Gary. 2022. Full moons and forking paths. *Significance*, 19(4), 32–35.

Smith, Gary, and Jay Cordes. 2019. *The 9 Pitfalls of Data Science*. Oxford: Oxford University Press.

Soares, Isa. 2017. The "fake news" machine: Inside a town gearing up for 2020. CNN, September 13.

Sozzi, Brian. 2020. Why stock market traders should be terrified of robots in the next decade. Yahoo/Finance, January 2.

Spinak, Ernesto. 2014. In the beginning it was just plagiarism—now its computer-generated fake papers as well. SciELO in Perspective, March 31.

Stempniak, Marty. 2021. Only 30% of radiologists currently using artificial intelligence as part of their practice. Radiology Business, April 21.

Sterling, T. D., Rosenbaum, W. L., & Weinkam, J. J. 1995. Publication decisions revisited: The effect of the outcome of statistical tests on the decision to publish and vice versa. *The American Statistician*, 49(1), 108–112.

Stodden, Victoria, Seiler, Jennifer, & Ma, Zhaokun. 2018. An empirical analysis of journal policy effectiveness for computational reproducibility. *Proceedings of the National Academy of Sciences*, 115(11), 2584–2589.

Subramanian, Samanth. 2017. Meet the Macedonian teens who mastered fake news and corrupted the US election. Wired, February 15.

Sullivan, Peter. 2021. Lancet report faults Trump for "avoidable" coronavirus deaths. The Hill, February 11.

Sydell, Laura. 2016. We tracked down a fake-news creator in the suburbs. Here's what we learned. All Tech Considered. NPR, November 23.

Synovitz, Ron, & Mitevska, Maria. 2020. "Fake news" sites in North Macedonia pose as American conservatives ahead of U.S. election. Radio Free Europe, October 22.

Theobald, Robert. 1963. *Free Men and Free Markets*. Garden City, NY: Anchor Books.

Thompson, Stuart A. 2022. How Trump Coins became an internet sensation. *New York Times*, January 28.

Tourianski, Julia. 2014. The declaration of bitcoin's independence. *Bitcoin Magazine*, May 14.

Townsend, Tess. 2016. The bizarre truth behind the biggest pro-Trump Facebook hoaxes. Inc., November 21.

Van der Zee, Tim, Anaya, Jordan, & Brown, Nicholas, J. L. 2017. Statistical heartburn: An attempt to digest four pizza publications from the Cornell Food and Brand Lab. *BMC Nutrition*, 3(54).

Van Noordan, Richard. 2014. Publishers withdraw more than 120 gibberish papers. *Nature*, February 25.

Van Noordan, Richard. 2021. Hundreds of gibberish papers still lurk in the scientific literature. *Nature*, May 27.

Vigna, Paul. 2019. Most bitcoin trading faked by unregulated exchanges, study finds. *Wall Street Journal*, March 22.

Vigna, Paul. 2019. Large bitcoin player manipulated price sharply higher, study says. *Wall Street Journal*, November 4.

Vosoughi, Soroush, Roy, Deb, & Aral, Sinan. 2018. The spread of true and false news online. *Science*, 359(6380), 1146–1151.

Wachter, Robert. 2017. *The Digital Doctor: Hope, Hype, and Harm at the Dawn of Medicine's Computer Age*. New York: McGraw-Hill Education.

Wagner, Rodd. 2016. The junk science of recognition by ratio. *Forbes*, July 13.

Wall Street Journal Staff. 2021. Facebook's documents about Instagram and teens. *Wall Street Journal*, September 29.

Wansink, Brian. 2016. The grad student who never said "No." *Healthier & Happier*, November 21.

Wansink, Brian. 2019. *Research Opportunities to Change Eating Behavior*.

Weimann, Gabriel. 2015. *Terrorism in Cyberspace: The Next Generation.* Washington, DC: Woodrow Wilson Center Press, 138.

Wells, Georgia, & Horwitz, Jeff. 2021. Facebook's effort to attract preteens goes beyond Instagram kids, documents show. *Wall Street Journal,* September 28.

Wells, Georgia, Horwitz, Jeff, & Glazer, Emily. 2021. How Facebook gobbled Mark Zuckerberg's bid to get America vaccinated. *Wall Street Journal,* September 17.

Wells, Georgia, Horwitz, Jeff, & Seetharaman, Deepa. 2021. Facebook knows Instagram is toxic for many teen girls, company documents show. *Wall Street Journal,* September 14.

Wicherts, Jelte. M., Borsboom, Denny, Kats, Judith, & Molenaar, Dylan. 2006. The poor availability of psychological research data for reanalysis. *American Psychologist,* 61(7), 726.

Wolff-Mann, Ethan. 2021. How to find bitcoin and other crypto asset "fundamentals": Goldman Sachs. YahooFinance, July 20.

Wynants, L., Van Calster, B., Collins, G. S., Riley, R. D., Heinze, G., Schuit, E., et al. 2020. Prediction models for diagnosis and prognosis of covid-19: systematic review and critical appraisal. *British Medical Journal,* 369.

Young, Stanley S., Bang, Heejung, and Oktay, Kutluk. 2009. Cereal-induced gender selection? Most likely a multiple testing false positive. *The Royal Society: Proceedings: Biological Sciences,* 276(1660), 1211–1212.

Zalesskaya, Yana. 2021. Meet your new A.I. best friend. *Fortune,* December.

Ziliak, ST, McCloskey, DN. (2004) Size matters: The standard error of regressions in the American Economic Review. *Journal of Socio-Economics,* 33(5): 527–546.

Zweig, Jason. 2018. When your investing robot has a mind of its own. *Wall Street Journal,* May 18.

INDEX

Please note that page references to Figures will be followed by the letter 'f', to Tables by the letter 't'.